Ashish Ghosh, Satchidananda Dehuri and Susmita Ghosh (Eds.)

Multi-Objective Evolutionary Algorithms for Knowledge Discovery from Databases

Studies in Computational Intelligence, Volume 98

Editor-in-chief
Prof. Janusz Kacprzyk
Systems Research Institute
Polish Academy of Sciences
ul. Newelska 6
01-447 Warsaw
Poland
E-mail: kacprzyk@ibspan.waw.pl

Further volumes of this series can be found on our homepage: springer.com

Vol. 72. Raymond S.T. Lee and Vincenzo Loia (Eds.)
Computation Intelligence for Agent-based Systems, 2007
ISBN 978-3-540-73175-7

Vol. 73. Petra Perner (Ed.)
Case-Based Reasoning on Images and Signals, 2008
ISBN 978-3-540-73178-8

Vol. 74. Robert Schaefer
Foundation of Global Genetic Optimization, 2007
ISBN 978-3-540-73191-7

Vol. 75. Crina Grosan, Ajith Abraham and Hisao Ishibuchi (Eds.)
Hybrid Evolutionary Algorithms, 2007
ISBN 978-3-540-73296-9

Vol. 76. Subhas Chandra Mukhopadhyay and Gourab Sen Gupta (Eds.)
Autonomous Robots and Agents, 2007
ISBN 978-3-540-73423-9

Vol. 77. Barbara Hammer and Pascal Hitzler (Eds.)
Perspectives of Neural-Symbolic Integration, 2007
ISBN 978-3-540-73953-1

Vol. 78. Costin Badica and Marcin Paprzycki (Eds.)
Intelligent and Distributed Computing, 2008
ISBN 978-3-540-74929-5

Vol. 79. Xing Cai and T.-C. Jim Yeh (Eds.)
Quantitative Information Fusion for Hydrological Sciences, 2008
ISBN 978-3-540-75383-4

Vol. 80. Joachim Diederich
Rule Extraction from Support Vector Machines, 2008
ISBN 978-3-540-75389-6

Vol. 81. K. Sridharan
Robotic Exploration and Landmark Determination, 2008
ISBN 978-3-540-75393-3

Vol. 82. Ajith Abraham, Crina Grosan and Witold Pedrycz (Eds.)
Engineering Evolutionary Intelligent Systems, 2008
ISBN 978-3-540-75395-7

Vol. 83. Bhanu Prasad and S.R.M. Prasanna (Eds.)
Speech, Audio, Image and Biomedical Signal Processing using Neural Networks, 2008
ISBN 978-3-540-75397-1

Vol. 84. Marek R. Ogiela and Ryszard Tadeusiewicz
Modern Computational Intelligence Methods for the Interpretation of Medical Images, 2008
ISBN 978-3-540-75399-5

Vol. 85. Arpad Kelemen, Ajith Abraham and Yulan Liang (Eds.)
Computational Intelligence in Medical Informatics, 2008
ISBN 978-3-540-75766-5

Vol. 86. Zbigniew Les and Mogdalena Les
Shape Understanding Systems, 2008
ISBN 978-3-540-75768-9

Vol. 87. Yuri Avramenko and Andrzej Kraslawski
Case Based Design, 2008
ISBN 978-3-540-75705-4

Vol. 88. Tina Yu, David Davis, Cem Baydar and Rajkumar Roy (Eds.)
Evolutionary Computation in Practice, 2008
ISBN 978-3-540-75770-2

Vol. 89. Ito Takayuki, Hattori Hiromitsu, Zhang Minjie and Matsuo Tokuro (Eds.)
Rational, Robust, Secure, 2008
ISBN 978-3-540-76281-2

Vol. 90. Simone Marinai and Hiromichi Fujisawa (Eds.)
Machine Learning in Document Analysis and Recognition, 2008
ISBN 978-3-540-76279-9

Vol. 91. Horst Bunke, Kandel Abraham and Last Mark (Eds.)
Applied Pattern Recognition, 2008
ISBN 978-3-540-76830-2

Vol. 92. Ang Yang, Yin Shan and Lam Thu Bui (Eds.)
Success in Evolutionary Computation, 2008
ISBN 978-3-540-76285-0

Vol. 93. Manolis Wallace, Marios Angelides and Phivos Mylonas (Eds.)
Advances in Semantic Media Adaptation and Personalization, 2008
ISBN 978-3-540-76359-8

Vol. 94. Arpad Kelemen, Ajith Abraham and Yuehui Chen (Eds.)
Computational Intelligence in Bioinformatics, 2008
ISBN 978-3-540-76802-9

Vol. 95. Radu Dogaru
Systematic Design for Emergence in Cellular Nonlinear Networks, 2008
ISBN 978-3-540-76800-5

Vol. 96. Aboul-Ella Hassanien, Ajith Abraham and Janusz Kacprzyk (Eds.)
Computational Intelligence in Multimedia Processing: Recent Advances, 2008
ISBN 978-3-540-76826-5

Vol. 98. Ashish Ghosh, Satchidananda Dehuri and Susmita Ghosh (Eds.)
Multi-Objective Evolutionary Algorithms for Knowledge Discovery from Databases, 2008
ISBN 978-3-540-77466-2

Ashish Ghosh
Satchidananda Dehuri
Susmita Ghosh
(Eds.)

Multi-Objective Evolutionary Algorithms for Knowledge Discovery from Databases

With 67 Figures and 17 Tables

Springer

Ashish Ghosh
Machine Intelligence Unit
and Center for Soft Computing Research
Indian Statistical Institute
203 B. T. Road
Kolkata 700 108
India
ash@isical.ac.in

Satchidananda Dehuri
Department of Information
and Communication Technology
F. M. University
Balasore 756 019
India
satchi.lapa@gmail.com

Susmita Ghosh
Department of Computer Science
and Engineering
Jadavpur University
Kolkata 700 032
India
susmitaghoshju@gmail.com

ISBN 978-3-540-77466-2 e-ISBN 978-3-540-77467-9

Studies in Computational Intelligence ISSN 1860-949X

Library of Congress Control Number: 2008921361

Cover design: Deblik, Berlin, Germany

Printed on acid-free paper

9 8 7 6 5 4 3 2 1

springer.com

To Our Parents

Preface

With the growth of information technology at unprecedented rate, we have witnessed a proliferation of different kinds of databases like biological, scientific, and commercial. Nowadays, it is fairly easy to create and customize a database tailored to our need. Nevertheless, as the number of records grows, it is not that easy to analyze and retrieve high level knowledge from the same database. There are not as many of-the-shelf solutions for data analysis as there are for database creation and management; furthermore, they are pretty hard to suit to our need.

Data Mining (DM) is the most commonly used name to describe such computational analysis of data and the results obtained must conform to several objectives such as accuracy, comprehensibility, interest for the user etc. Though there are many sophisticated techniques developed by various interdisciplinary fields only a few of them are well equipped to handle these multi-criteria issues of DM. Therefore, the DM issues have attracted considerable attention of the well established multi-objective genetic algorithm community to optimize the objectives in the tasks of DM.

The present volume provides a collection of seven articles containing new and high quality research results demonstrating the significance of Multi-objective Evolutionary Algorithms (MOEA) for data mining tasks in Knowledge Discovery from Databases (KDD). These articles are written by leading experts around the world. It is shown how the different MOEAs can be utilized, both in individual and integrated manner, in various ways to efficiently mine data from large databases.

Chapter 1, by Dehuri et al., combines activities from three different areas of active research: knowledge discovery from databases, genetic algorithms and multi-objective optimization. The goal of this chapter is to identify the objectives those are implicitly/explicitly associated with the tasks of KDD like pre-processing, data mining, and post processing and then discuss how MOEA can be used to optimize them.

Chapter 2 contributed by Landa-Becerra et al. presents a survey of techniques used to incorporate knowledge into evolutionary algorithms, with special emphasis on multi-objective optimization. They focus on two main groups of techniques: techniques which incorporate knowledge into fitness evaluation, and those which incorporate knowledge in the initialization process and the operators of an evolutionary algorithm. Several methods, representative of each of these groups, are briefly discussed together with some examples found in the specialized literature. In the last part of the chapter, the authors provided some research ideas that are worth exploring in future by researchers interested in this topic.

Classification rule mining is one of the fundamental tasks of data mining. Ishibuchi et al. has solved this problem using evolutionary multi-objective optimization algorithms and their work is included in Chapter 3. In the field of classification rule mining, classifiers are designed through the following two phases: rule discovery and rule selection. In rule discovery phase, a large number of classification rules are extracted from training data. This phase is based on two rule evaluation criteria: support and confidence. An association rule mining technique such as Apriori algorithm is usually used to extract classification rules satisfying pre-specified threshold values of the minimum support and confidence. In the second phase a small number of rules are selected from the extracted rules to design an accurate and compact classifier. In this chapter, first the authors explained the above-mentioned two phases in classification rule mining. Next they described how to find the Pareto-optimal rules and Pareto-optimal rule sets. Then they discussed evolutionary multi-objective rule selection as a post processing procedure.

Chapter 4, written by Jin et al., showed that rule extraction from neural networks is a powerful tool for knowledge discovery from data. In order to facilitate rule extraction, trained neural networks are often pruned so that the extracted rules are understandable to human users. This chapter presents a method for extracting interpretable rules from neural networks that are generated using an evolutionary multi-objective algorithm. In the algorithm, the accuracy on the training data and the complexity of the neural networks are minimized simultaneously. Since there is a tradeoff between accuracy and complexity, a number of Pareto-optimal neural networks, instead of a single optimal neural network, are obtained. They showed that the Pareto-optimal networks, with a minimal degree of complexity, are often interpretable as they can extract understandable logic rules. Finally they have verified their approach using two benchmark problems.

Alcala et al. contributed Chapter 5 which deals with the usefulness of MOEAs for getting compact fuzzy rule based systems (FRBSs) under parameter tuning and rule selection. This contribution briefly reviews the state-of-the-art of this topic and presents an approach to prove the ability of multi-objective genetic algorithms for obtaining compact fuzzy rule based systems under rule selection and parameter tuning, i.e., to obtain linguistic models with improved accuracy and minimum number of rules.

Chapter 6, contributed by Setzkorn, deals with the details of three MOEAs to solve different data mining problems. The first approach is used to induce fuzzy classification rule systems and the other two are used for survival analysis problems. Till now, many evolutionary approaches use accuracy to measure the fitness of the model to the data. This is inappropriate when the misclassifications costs and class prior probabilities are unknown, which is often the case in practice. Hence, the author uses a measure called the area under the receiver-operating characteristic curve

(AUC) that does not have these problems. The author also deploys a self-adaptation mechanism to reduce the number of free parameters and uses the state-of-the-art multi-objective evolutionary algorithm components.

Chapter 7, written by Murty et al., discusses the role of evolutionary algorithms (EAs) in clustering. In this context, they have pointed out that most of the GA-based clustering algorithms are applied on the data sets with small number of patterns and or features. Hence, to cope up with large size and high dimensional data sets, they proposed a GA based algorithm, OCFTBA, employing the cluster feature tree (CF-tree) data structure. It scales up well with the number of patterns and features. They have also suggested that clustering for the large scale data can be formulated as a multi-objective problem and solving them using GAs will be very interesting with a flavor of good set of Pareto optimal solutions.

Kolkata, Balasore
October 2007

Ashish Ghosh
Satchidananda Dehuri
Susmita Ghosh

Contents

List of Contributors

R. Alcalá
Department of Computer Science and Artificial Intelligence
University of Granada
E-18071 - Granada, Spain
alcala@decsai.ugr.es

J. Alcalá-Fdez
Department of Computer Science and Artificial Intelligence
University of Granada
E-18071 - Granada, Spain
jalcala@decsai.ugr.es

Chiranjib Bhattacharyya
Department of Computer Science and Automation
Indian Institute of Science
Bangalore 560012, India
chiru@csa.iisc.ernet.in

Carlos A. Coello Coello
CINVESTAV-IPN (Evolutionary Computation Group)
Departamento de Computación
Av. IPN No. 2508
Col. San Pedro Zacatenco, México
D. F. 07360, Mexico
ccoello@cs.cinvestav.mx

Satchidananda Dehuri
Department of Information and Communication Technology
Fakir Mohan University
Vyasa Vihar
Balasore 756019, India
satchi.lapa@gmail.com

M. J. Gacto
Department of Computer Science and Artificial Intelligence
University of Granada
E-18071 - Granada, Spain
mjgacto@ugr.es

Ashish Ghosh
Machine Intelligence Unit
Indian Statistical Institute
203 B.T. Road, Kolkata 700108, India
ash@isical.ac.in

Susmita Ghosh
Department of Computer Science and Engineering, Jadavpur University
Kolkata 700032, India
susmitaghoshju@gmail.com

F. Herrera
Department of Computer Science and Artificial Intelligence
University of Granada
E-18071 - Granada, Spain
herrera@decsai.ugr.es

Hisao Ishibuchi
Department of Computer Science and Intelligent Systems
Graduate School of Engineering
Osaka Prefecture University
1-1 Gakuen-cho, Naka-ku, Sakai
Osaka 599-8531, Japan
hisaoi@cs.osakafu-u.ac.jp

Yaochu Jin
Honda Research Institute Europe
63073 Offenbach/Main, Germany
yaochu.jin@honda-ri.de

Edgar Körner
Honda Research Institute Europe
63073 Offenbach/Main, Germany

Isao Kuwajima
Department of Computer Science and Intelligent Systems
Graduate School of Engineering
Osaka Prefecture University
1-1 Gakuen-cho, Naka-ku, Sakai
Osaka 599-8531, Japan
kuwajima@cs.osakafu-u.ac.jp

Ricardo Landa-Becerra
CINVESTAV-IPN (Evolutionary Computation Group)
Departamento de Computación
Av. IPN No. 2508
Col. San Pedro Zacatenco, México
D. F. 07360, Mexico
rlanda@computacion.cs.cinvestav.mx

M. N. Murty
Department of Computer Science and Automation
Indian Institute of Science
Bangalore 560012, India
mnm@csa.iisc.ernet.in

Yusuke Nojima
Department of Computer Science and Intelligent Systems
Graduate School of Engineering
Osaka Prefecture University
1-1 Gakuen-cho, Naka-ku, Sakai
Osaka 599-8531, Japan
nojima@cs.osakafu-u.ac.jp

Babaria Rashmin
Department of Computer Science and Automation
Indian Institute of Science
Bangalore 560012, India
rashmin@csa.iisc.ernet.in

Luis V. Santana-Quintero
CINVESTAV-IPN (Evolutionary Computation Group)
Departamento de Computación
Av. IPN No. 2508
Col. San Pedro Zacatenco, México
D. F. 07360, Mexico
lvspenny@hotmail.com

Bernhard Sendhoff
Honda Research Institute Europe
63073 Offenbach/Main, Germany

Christian Setzkorn
National Center for Zoonosis Research
University of Liverpool, UK
christian@setzkorn.eu

1

Genetic Algorithm for Optimization of Multiple Objectives in Knowledge Discovery from Large Databases

Satchidananda Dehuri[1], Susmita Ghosh[2], Ashish Ghosh[3]

[1] Machine Intelligence Unit and Center for Soft Computing Research, Indian Statistical Institute, 203 B.T. Road, Kolkata 700108, INDIA. ash@isical.ac.in
[2] Department of Information and Communication Technology, Fakir Mohan University, Vyasa Vihar, Balasore 756019, INDIA. satchi.lapa@gmail.com
[3] Department of Computer Science and Engineering, Jadavpur University, Kolkata 700032, INDIA. susmitaghoshju@gmail.com

Summary. Knowledge discovery in databases (KDD) is increasingly being accepted as a viable tool for collecting, analyzing, and making decision from massive data sets. Though many sophisticated techniques are developed by various interdisciplinary fields, only few of them are well equipped to handle multi-criteria issues of KDD. It seems to provide a new frontier of research directions. The KDD issues like feature selection, instance selection, rule mining and clustering involves simultaneous optimization of several (possibly conflicting) objectives. Further, considering a single criterion, as with the existing soft computing techniques like evolutionary algorithms (EA), neural network (NN), and particle swarm optimization (PSO) are not up to the mark. Therefore, the KDD issues have attracted considerable attention of the well established multi-objective genetic algorithms to optimize the identifiable objectives in KDD.

1.1 Introduction

This chapter combines works from three different areas of active research: knowledge discovery in databases, genetic algorithms and multi-objective optimization. The goal of this chapter is to identify the objectives those are implicitly/explicitly associated with the steps involved in KDD process like pre-processing, data mining, and post processing; and then discuss how multi-objective genetic algorithms (MOGA) can be used to optimize these objectives.

Knowledge discovery in databases creates the context for developing new generation computational techniques and tools to support the extraction of useful knowledge from the rapidly growing volumes of data (16). The KDD process such as data pre-processing and post processing, in addition to data mining-application of specific algorithms for extracting patterns (models) from data, ensures that useful knowledge is derived from the data. The pre-processing steps of KDD such as feature selection (41), instance selection (20), involve multiple and often conflicting criteria, which is

S. Dehuri et al.: *Genetic Algorithm for Optimization of Multiple Objectives in Knowledge Discovery from Large Databases*, Studies in Computational Intelligence (SCI) **98**, 1–22 (2008)
www.springerlink.com

a non-trivial problem of optimization. Similarly, in the case of rule mining and clustering it is very difficult to get an acceptable model with multi-criteria. Even though many sophisticated techniques are developed to find out models by various interdisciplinary fields but none of them are well equipped to handle multi-criteria issues of KDD. Henceforth it is a challenging issue to find out the multiple criteria in each step of KDD and optimize in the context of classification, association rule mining and clustering. Let us discuss briefly what are KDD, GAs (21), and multi-objective genetic algorithms (MOGAs) (1) and how they help us to reach the target.

1.1.1 An Introduction to KDD

During recent years, the amount of data in the world is doubling every twenty months (18). The reason is that due to rapid growth of data collection technology such as scanners, bar code reader for all commercial products, and the computerization of many business (e.g., credit card purchases) and government transactions (e.g. tax returns) and sensors in scientific and industrial domains (16) (e.g., from remote sensors or from space satellites). Furthermore, the advances in data storage technology, such as faster, higher capacity, and cheaper storage devices allowed us to transform this data deluge into mountains of stored data. In scientific endeavors, data represents observations carefully collected about some phenomenon under study. In business, data captures information about critical markets, competitors, and customers. In manufacturing, data captures performance and optimization opportunities, as well as the keys to improve processes and troubleshooting problems. But one common question arises why do people store this vast collection of data? Raw data is rarely of direct benefit. Its true value is reflected in the ability to extract useful information for decision support or exploration and understanding of the phenomena governing the data source. Traditionally, analysis was strictly a manual process. One or more analysts would become intimately familiar with the data and with the help of statistical techniques they would provide summaries and generate reports. In effect, the analysts acted as sophisticated query processors. However, such an approach rapidly breaks down as the quantity of data grows and the number of dimensions increases. A few examples of large data in real life applications are as follows. In the business world, one of the largest databases in the world is created by Wal-Mart (a U. S. retailer), which handles over 20 million transactions a day (21). There are huge scientific databases as well. The human genome database project (23) has collected gigabytes of data on the human genetic code and much more is expected. A database housing a sky object catalog from a major astronomy sky survey (24; 25) consists of billions of entries with raw image data sizes measured in terabytes. The NASA Earth Observing System (EOS) of orbiting satellites and other space borne instruments is projected to generate about 50 gigabytes of remotely sensed image data per hour. Such volumes of data clearly overwhelm the traditional manual methods of data analysis. All these have prompted the need for new generation tools and techniques with the ability to intelligently and automatically assist humans in analyzing the mountains of data for nuggets of useful knowledge. These techniques and tools are the subject of the emerging field of knowledge discovery in databases (KDD). A number of successful applications of KDD technology have been reported in (16; 42; 43).

The KDD Process: The knowledge discovery in databases is the non-trivial process of identifying *valid, novel, potentially useful*, and *ultimately understandable patterns* in data (16). The overall KDD process is shown in Figure 1.1. It is interactive and iterative, involving a large number of steps with many decisions being made by the user. Brachman & Anand (17) gave a practical view of the KDD process emphasizing the interactive nature of the process (16). Here we broadly outline some of its basic steps:

1. Developing an understanding of the application domain: the relevant prior knowledge, and the goals of the end-user.
2. Creating a target data set: selecting a data set, or focusing on a subset of variables or data samples, on which discovery is to be performed.
3. Data cleaning and pre-processing: basic operations such as the removal of noise or outliers, if appropriate, collecting the necessary information to model or account for noise, deciding on strategies for handling missing data fields, accounting for time sequence information and known changes.
4. Data reduction and projection: finding useful features to represent the data depending on the goal of the task. Using dimensionality reduction or transformation methods to reduce the effective number of variables under consideration or to find invariant representations for the data.
5. Choosing the data mining task: deciding whether the goal of the KDD process is classification, regression, clustering etc.
6. Choosing the data mining algorithm(s): selecting method(s) to be used for searching for patterns in the data. This includes deciding which models and parameters may be appropriate (e.g., models for categorical data are different than models on vectors over the real) and matching a particular data mining method with the overall criteria of the KDD process.
7. Data mining: searching for patterns of interest in a particular representational form or a set of such representations: classification rules, regression, clustering, and so forth. The user can significantly aid the data mining method by correctly performing the preceding steps.
8. Interpreting mined patterns, possible return to any steps from 1-7 for further iteration.
9. Consolidating discovered knowledge: incorporating this knowledge into the performance system, or simply documenting it and reporting it to interested parties. This also includes checking for and resolving potential conflicts with previously believed (or extracted) knowledge.

1.1.2 Genetic Algorithms

As the problems are becoming larger and more complex, researchers are turning towards heuristic techniques to complement existing approaches. Here we introduce

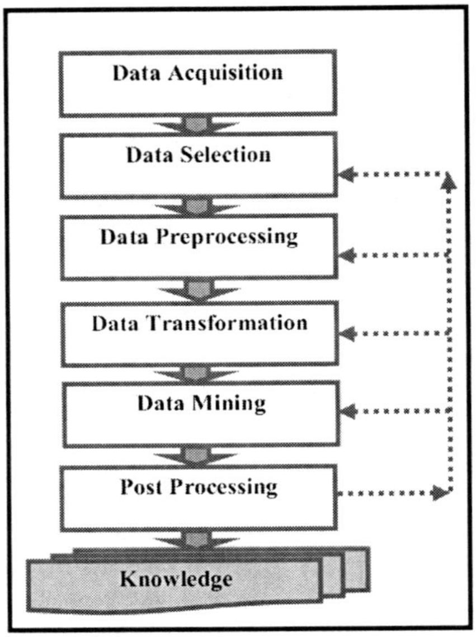

Fig. 1.1. Basics steps of KDD process

genetic algorithms, as our aim is to examine the role of genetic algorithms to solve many of the multi-objective optimization problems involved in KDD process.

A variety of evolutionary computational models have been proposed and studied and are referred to as EAs (6; 26). An EA uses computational models of evolutionary processes as key elements in the design and implementation of computer-based problem solving systems. There have been many well-defined EAs, which have served as the basis for much of the activity in the field: genetic algorithms (GAs)(2), evolution strategies (3; 51), genetic programming (GP) (5) evolutionary programming (6; 14; 15), cultural evolution (4), co-evolution (44) etc. An EA maintains a population of trial solutions, imposes changes to these solutions, and incorporates an environmental pressure called selection to determine which ones are going to be maintained in future generations and which will be discarded from the pool of trials. There are some important differences between the existing EAs. GAs emphasize models of genetic operators as observed in nature, such as crossover (recombination) and mutation, and apply these to abstracted chromosomes with different representation schemes according to the problem being solved. Evolution strategies and evolutionary programming only apply to real-valued problems and emphasize mutational transformations that maintain the behavioral linkage between each parent and its offspring. As regards to GP, it constitutes a variant of GAs, based on evolving structures encoding programs such as expression trees. Apart from adapting the crossover and mutation operators to deal with the specific coding scheme considered, the remaining algorithmic components remain the same. Unlike standard EAs, which are unbiased, using little or no domain knowledge to guide the search process, cultural evolution (CE) based upon the principles of human social evolution, was developed as an approach to bias the search process with prior knowledge about the domain. Cultural evolution algorithms model two search spaces, namely the *population space* and the *belief space*. The belief space models the cultural information about the population. Both the population and the beliefs evolve with both spaces influencing one another. Coevolution is the complementary evolution of closely associated species, where there is an inverse fitness interaction between the two species. A win for one species means a failure for the other. To survive, the losing species adapt to counter the winning species in order to become the new winner. An alternative coevolutionary process is symbiosis, in which the species cooperate instead of competing. In this case a success in one species improves the survival strength of the other species. In standard EAs, evolution is usually viewed as if the population attempts to adapt in a fixed physical environment. In contrast, coevolutionary algorithms realize that in natural evolution the physical environment is influenced by other independently acting biological populations. As this chapter is trying to explore the wide applicability of genetic algorithms in KDD and its de-facto standard for solving multi-objective problems, it is our duty to discuss briefly about genetic algorithm.

Genetic algorithms are probabilistic search algorithms characterized by the fact that a number N of potential solutions (called individuals $I_k \in \Omega$, where Ω represents the space of all possible individuals) of the optimization problem simultaneously samples the search space. This population $P = \{I_1, I_2, ..., I_N\}$ is modified according to the natural evolutionary process: after initialization, selection $S : I^N \rightarrow I^N$ and recombination $\Re : I^N \rightarrow I^N$ are executed in a loop until some termination criterion is reached. Each run of the loop is called a generation and $P(t)$ denotes the population at generation t. The selection operator is intended to improve the average quality of the population by giving individuals of higher quality a higher probability to be copied into the next generation. Selection, thereby, focuses on the search of promising regions in the search space. The quality of an individual is measured by a fitness function $f : P \rightarrow R$. Recombination changes the genetic material in the population either by crossover or by mutation in order to obtain new points in the search space. Figure 1.2 depicts the steps that are performed in GAs.

Genetic algorithms have often been criticized when applied to data mining, because of the amount of computational resources they require, and because they are unable to find optimal solutions deterministically due to their stochastic nature. However, we believe that genetic algorithms have a lot of potential in data mining for the following reasons: they can handle attribute interaction and are well suited for solving multi-objective problems. Also, in complex domains, it is not feasible to find all the optimal solutions of a problem. With the increasing speed of processors and the scope of parallelization, we believe the computational cost can be justified.

1.1.3 Multi-objective Genetic Algorithms

As a KDD process implicitly/explicitly involves many criteria to be optimized simultaneously, solution to such problems are usually computed by combining them

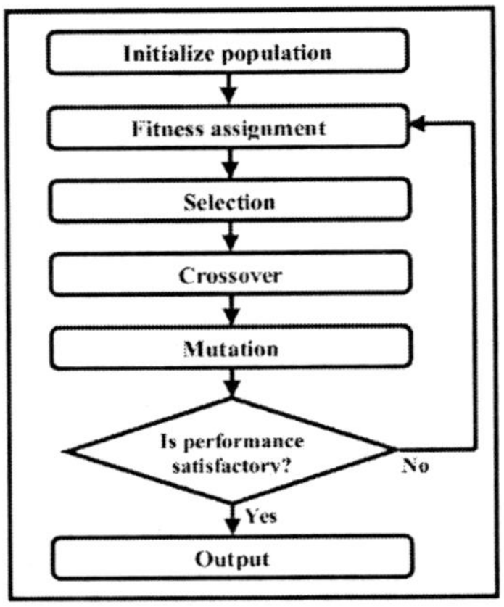

Fig. 1.2. Flow diagram of genetic algorithm

into a single criterion optimization problem. But the resulting solution to the single objective optimization problem is usually subjective to the parameter settings. Moreover, since usually a classical optimization method is used, only one solution (hopefully, a Pareto optimal solution) can be found in one simulation. Thus in order to find multiple Pareto optimal solutions, genetic algorithms are the best choice, because they deal with a population of solutions. It allows finding an entire set of Pareto optimal solutions in a single run of the algorithm. In addition to this, genetic algorithms are susceptible to the shape or continuity of the Pareto front. There are many multi-objective problems requiring simultaneous optimization of several competing objectives. Formally it can be stated as follows.

We want to find $\overrightarrow{x} = \{x_1, x_2, x_3, ..., x_n\}$ which maximizes the values of p objective functions $F(\overrightarrow{x}) = \langle f_1(\overrightarrow{x}), f_2(\overrightarrow{x}), f_3(\overrightarrow{x}), ..., f_p(\overrightarrow{x})\rangle$within a feasible domain Ω. Generally the answer is not a single solution but a family of solutions called a *Pareto-optimal set.*

Definitions: A vector $\overrightarrow{u} = \langle u_1, u_2, ..., u_p\rangle$ is said to dominate another vector $\overrightarrow{v} = \langle v_1, v_2, ..., v_p\rangle$ iff $\overrightarrow{u}$ is partially greater than $\overrightarrow{v} i.e. \forall i \in \{1, 2, 3, .., p\}, u_i \geq v_i \& \exists i \in \{1, 2, 3, .., p\} : u_i \succ v_i$.

A solution $x \in \Omega$ is said to be Pareto-optimal with respect to Ω iff there is no $x' \in \Omega$ for which $\overrightarrow{v} = F(x') = \langle f_1(x'), f_2(x'), ..., f_p(x')\rangle$ dominates $\overrightarrow{u} = F(x) = \langle f_1(x), f_2(x), ..., f_p(x)\rangle$.

For a given multi-objective problem $F(x)$, the Pareto-optimal set P_s is defined as:

$P_s = \langle x \in \Omega | \neg\exists : F(x') \geq F(x)\rangle$.

For a given multi-objective problem $F(x)$ and Pareto optimal set P_s, the Pareto front P_f is defined as:

$P_f = \langle \overrightarrow{u} = F(x) = (f_1(x), f_2(x), ..., f_p(x) | x \in P_s)\rangle$.

Optimization methods generally try to find a given number of Pareto-optimal solutions which are uniformly distributed in the Pareto-optimal set. Such solutions provide the decision maker sufficient insight into the problem to make the final decision. Pareto optimal front of a set of solutions generated from two conflicting objectives

$$f_1(x,y) = \frac{1}{(x^2+y^2+1)}$$

and

$$f_2(x,y) = x^2 + 3.y^2 + 1, -3 \leq x, y \leq +3$$

is illustrated in Figure 1.3.

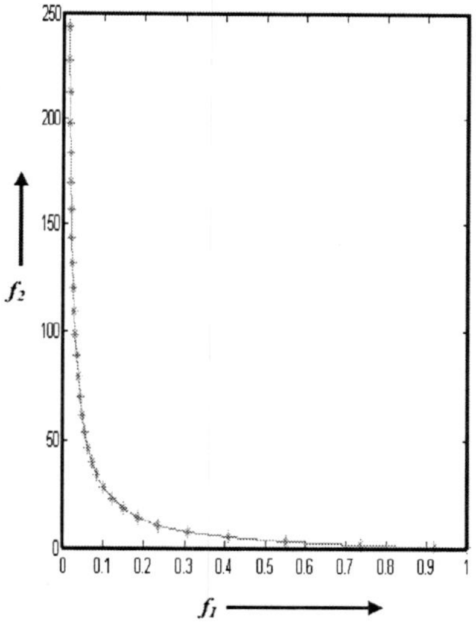

Fig. 1.3. Pareto optimal front

To cope with this multi-objective problems one can review three very different approaches, namely: i) transforming the original multi-objective problem into a single objective problem by using a weighted function, ii) the lexicographical approach, where the objectives are ranked in order of priority, and iii) the Pareto approach which consists of as many non-dominated solutions as possible and returning the set of Pareto front to the user. The general conclusions are that the weighted formula approach, which is by far the most used in the data mining literature, is an ad-hoc approach for multi-objective optimization, whereas the lexicographic and the Pareto approaches are more principled approaches, and therefore deserved more attention from the data mining community. The three broad categories to cope with multi-objective problems are described as follows:

Weighted Sum Approach

In the data mining literature, by far, the most widely used approach to cope with a multi-objective problem consists of transforming it into a single objective one. This is typically done by multiplying a numerical weight to each objective (evaluation criterion) and then combining the values of the weighted criterion into a single value. That is, the fitness value 'F' of a given candidate rule is typically given by one of the following formulas:

$$F = w_1 \cdot f_1 + w_2 \cdot f_2 + ... + w_n \cdot f_n, \tag{1.1}$$

where $w_i, i = 1, 2, 3...n$, denotes the weight assigned to criterion i and n is the number of evaluation criteria.

The advantage of this method is its simplicity. However it has several drawbacks. The most obvious problem with weighted sum approach is that, in general, the setting of the weights in these formulas is ad-hoc, based either on an intuition of the user about the relative importance of different quality criteria or on trial and error experimentation with different weight values. Hence the values of these weights are determined empirically. Another problem with these weights is that, once a formula with precise values of weights has been defined and given to a data mining algorithm, the data mining algorithm will be effectively trying to find the best rule for that particular settings of weights, missing the opportunity to find other rules that might be actually more interesting to the user, representing a better trade-off between different quality criteria. In particular, weighted formulas involving a linear combination of different quality criteria have the limitation that they cannot find solutions in a non-convex region of the Pareto front. For now, to see the limitation of linear combinations of different criteria, consider a hypothetical scenario where we have to select the best rule to the position of data miner in a company, taking into account two criteria f_1 and f_2- say, their amount of knowledge about data mining and machine learning, measured by a test and/or a detailed technical interview. Suppose the first candidate's scores are $f_1 = 9.5$ and $f_2 = 5$; the second candidate's scores are $f_1 = 7$ and $f_2 = 7$; and the third candidate's scores are $f_1 = 5$ and $f_2 = 9.5$. The choice of the best candidate should depend, of course, on the relative importance assigned to the two criteria by the employer. It is interesting to note, however, that although it is trivial to think of weights for criteria f_1 and f_2 that would make the first or third candidate the winner, it is actually impossible to choose weights for f_1 and f_2 so that second candidate would be winner-assuming that the weighted formula is a linear combination of weights. Intuitively, however, the second candidate might be the favorite candidate of many employers, since she/he is the only one to have a good knowledge about both data mining and machine learning. Let us now turn to the problem of mixing non-commensurable criteria in a weighted formula evaluating a candidate rule. In particular, let us consider the problem of mixing accuracy and comprehensibility (simplicity) measures into the same formula, since these are probably the two-rule quality criteria most used in data mining. Clearly, accuracy and comprehensibility are two very different, non-commensurable criteria to evaluate the quality of a rule. Actually, comprehensibility is an inherently subjective, user dependent-criterion. Even if we replace the semantic notion of comprehensibility by a syntactic measure of simplicity such as rule size, as it is usually done when evaluating comprehensibility, the resulting measure of simplicity is still non-commensurable with a measure of accuracy. The crucial problem is not that these two criteria have different units of measurement (which can be, to some extent, reasonably solved by normalization, as discussed earlier), but rather they represent very different aspects of a rule's quality. In principle, it does not make sense to add/subtract accuracy and simplicity, and the meaning of an indicator multiplying/dividing these two quality criteria are questionable. A more meaningful approach is to recognize that accuracy and simplicity are two very different quality criteria and treat them separately without mixing them in the same formula.

Lexicographic Approach

The basic idea of this approach is to assign different priorities to different objectives, and then focus on optimizing the objectives in their order of priority. Hence, when two or more candidate rules are compared with each other to choose the best one, the first thing to do is to compare their performance measure for the highest priority objective. If one candidate rule is significantly better than the other with respect to that objective, the former is chosen. Otherwise the performance measure of the two candidate models is compared with respect to the second objective. Again if one candidate rule is significantly better than the other with respect to that objective, the former is chosen, otherwise the performance measure of the two-candidate rule is compared with respect to the third criterion. The process is repeated until one finds a clear winner or until one has used all the criteria. In the latter case, if there was no clear winner, one can simply select the model optimizing the highest priority objective.

The lexicographic approach has one important advantage over the weighted sum approach: the former avoids the problem of mixing non-commensurable criteria in the same formula. Indeed, the lexicographic approach treats each of the criteria separately, recognizing that each criterion measures a different aspect of quality of a candidate solution. As a result, the lexicographic approach avoids the drawbacks associated with the weighted sum approach such as the problem of fixing weights. In addition, although the lexicographic approach is somewhat more complex than the weighted-sum-formula approach, the former can still be considered conceptually simple and easy to use.

The lexicographic approach usually requires one to specify a tolerance threshold for each criterion. It is not trivial how to specify these thresholds in a principled manner. A commonplace approach is to use a statistics-oriented procedure, e.g., standard deviation-based thresholds, which allows us to reject a null hypothesis of insignificant difference between two objective values with a certain degree of confidence. This specification still has a certain degree of arbitrariness, since any high-value such as 95% or 99% could be used. Of course one can always ask the user to specify the thresholds or any other parameter, but this introduces some arbitrariness and subjectiveness in the lexicographic approach- analogous to the usually arbitrary, subjective specification of weights for different criteria in the weighted formula approach. Hence after analyzing the pros and cons of both these methods no one seems to be much suitable for rule mining problems associated with multiple objectives. Therefore, we study an alternative method called Pareto approach.

Pareto Approach

The basic idea of Pareto approach is that, instead of transforming a multi-objective one into a single objective problem and then solving it by a genetic algorithm, one should use multi-objective genetic algorithm directly. One should adapt the algorithm to the problem being solved, rather than the other way around. In any case, this intuition needs to be presented in more formal terms. Let us start with a definition of Pareto dominance. A solution x_1 is said to dominate a solution x_2 iff x_1 is strictly better than x_2 with respect to at least one of the criteria (objective) being optimized and x_1 is not worse than x_2 with respect to all the criteria being optimized. To solve this kind of mining problem by multi-objective genetic algorithm, the first task is to represent the possible rules as individuals known as individual representation. Second task is to define the fitness function and then genetic materials. A lot of multi-objective GAs (MOGAs)(1; 27) have been suggested in the literature.

Basically, a MOGA is characterized by its fitness assignment and diversity maintenance strategy. In fitness assignment, most MOGAs fall into two categories, non-Pareto and Pareto-based. Non-Pareto methods use the objective values as the fitness value to decide an individual's survival. Schaffer's VEGA (54) is such a method. The Predator-prey approach (28) is another one, where some randomly walking predators will kill a prey or let it survive according to the prey's value in one objective. In contrast, Pareto based methods measure individuals' fitness according to their dominance property. The non-dominated individuals in the population are regarded as fittest regardless of their single objective values. Since Pareto-based approaches give emphasis on the dominance nature of multi-objective problems, their performance is reported to be better.

Diversity maintenance strategy is another characteristic of MOGAs. It works by keeping the solutions uniformly distributed in the Pareto-optimal set, instead of gathering solutions in a small region only. Fitness sharing (29), which reduces the fitness of an individual if there are some other candidates nearby, is one of the most renowned techniques. Restricted mating, where mating is permitted only when the distance between two parents are large enough, is another technique. More recently, some parameter free techniques were suggested. The techniques used in SPEA (28) and NSGA-II (29) are two examples of such techniques. PAES (30), SPEA (28), and NSGA-II (29) are representatives of current MOGAs. They all adopt Pareto-based fitness assignment strategy and implement elitism, an experimentally verified technique known to enhance performance. A good comprehensive study of MOGA can be found in (27).

1.2 Problem Identification

In this section we discuss the objectives, that are implicitly/explicitly present in different steps of KDD process.

1.2.1 Multiple Criteria Problems in KDD

The multiple criteria in KDD is categorized into following three types:

Multiple Criteria in Pre-processing

In this step our basic objective is the selection of relevant features and a compact reference set called instance selection. In feature selection (31) the objective is to select a small number of relevant features with a high classification accuracy i.e. two primary objectives such as minimize the set of features and maximize the classification accuracy. Similarly, in the case of instance selection, the job is to minimize the reference set with an objective of high classification accuracy. Further we can combine the whole idea to a single problem, which is defined as minimize the number of instances and selected features and maximize the classification accuracy.

Mathematically, let us assume that m labeled instances $X = \{x_1, x_2, x_3, ..., x_m\}$ are taken from c classes. We also denote the set of given n features as $F = \{f_1, f_2, ..., f_n\}$. Let D and A be the set of selected instances and the set of selected features, respectively, where $D \subseteq X$ and $A \subseteq F$. We denote the reference set as $S = \langle D, A \rangle$. Now our problem is formulated as follows: *Minimize* $|D|$, *Minimize* $|A|$, and Maximize *Performance(S)*.

Multiple Criteria in Rule Mining

The multi-criteria problem in rule mining are categorized into two types: classification rule mining and association rule mining.

Classification Rule Mining: In classification rule mining the discovered rules should have (a) *high predictive accuracy*, (b) *comprehensibility* and (c) *interestingness.*

Comprehensibility Metric: There are various ways to quantitatively measure rule comprehensibility. A standard way of measuring comprehensibility is to count the number of rules and the number of conditions in these rules. If these numbers increase then the comprehensibility decrease. For instance, if a rule R has at most M conditions, the comprehensibility of a rule $C(R)$ can be defined as:

$$C(R) = M - (number\ of\ conditions\ (R)) \tag{1.2}$$

i.e. if the number of conditions in the rule antecedent part increases then the comprehensibility decreases.

Predictive Accuracy: Rules are of the form IF A1 and A2 THEN C. The antecedent part of the rule is a conjunction of conditions. A very simple way to measure the predictive accuracy of a rule is

$$Pred\_Accuracy(R) = \frac{|A\&C|}{|A|} \tag{1.3}$$

where $|A|$ is the number of instances satisfying all the conditions in the antecedent A and $|A\&C|$ is the number of examples that satisfy both the antecedent A and the consequent C. Intuitively, this metric measures predictive accuracy in terms of how many cases both antecedent and consequent hold out of all cases where the antecedent holds. However, it is quite prone to overfitting, because a rule covering small number of instances could have a high value even though such a rule would not be able to generalize the unseen data during training. An alternative measure of predictive accuracy of the rule is the number of correctly classified test instances divided by the total number (correctly classified + wrongly classified) of test instances. Although this method is widely used, it has a disadvantage of unbalanced class distribution (7).

Hence to avoid these limitations the following measure of predictive accuracy is taken into consideration and is discussed in more detail (7). A confusion matrix can summarize the performance of a classification rule with respect to predictive accuracy.

Let us consider the simplest case, where there are only two classes to be predicted, referred to as the class C and the class $C\prime$. In this case the confusion matrix will be a 2 × 2 matrix and is illustrated in Table 1.1.

Table 1.1. A 2 × 2 confusion matrix

		Actual Class	
Predicted		C	$C\prime$
Class	C	L_{CC}	$L_{CC\prime}$
	$C\prime$	$L_{C\prime C}$	$L_{C\prime C\prime}$

In Table 1.1, C denotes the class generated by a rule, and all other classes are considered as $C\prime$. The labels in each quadrant of the matrix have the following meaning:

- L_{CC}= Number of instances satisfying A and having class C.
- $L_{CC\prime}$= Number of instances satisfying A and having class $C\prime$.
- $L_{C\prime C}$= Number of instances not satisfying A but having class C.
- $L_{C\prime C\prime}$= Number of instances not satisfying A and having class $C\prime$.

Intuitively, the higher the values of the diagonal elements and lower the values of other elements, better is the corresponding classification rule. This matrix can also work for a 'm' class problem: when there are more than two classes one can still work with the assumption that algorithm evaluates one rule at a time and the class C predicted by the rule is considered as the C class, and all other classes are simply considered as $C\prime$ classes. Given the values of $L_{CC}, L_{CC\prime}, L_{C\prime C}$ and $L_{C\prime C\prime}$ as discussed above, the predictive accuracy is defined as

$$P(R) = \frac{L_{CC} \times L_{C\prime C\prime}}{(L_{CC} + L_{CC\prime}) \times (L_{CC} + L_{C\prime C})}, \tag{1.4}$$

where $0 \leq P(R) \leq 1$.

Interestingness: The computation of the degree of interestingness of a rule, in turn, consists of two terms. One of them refers to the antecedent of the rule and the other to the consequent. The degree of interestingness of the rule antecedent is calculated by an information-theoretical measure, which is a normalized version of the measure proposed in (45). Initially, as a pre-processing step, the algorithm calculates the information gain of each attribute (InfoGain) (8). Then the degree of interestingness (RInt) of the rule antecedent is given by:

$$RInt = 1 - \frac{\frac{\sum_{i=1}^{n-1} InfoGain(A_i)}{n-1}}{\ln(|dom(G)|)}, \tag{1.5}$$

where 'n' is the number of attributes (features) in the antecedent and $(|dom(G)|)$ is the domain cardinality (i.e. the number of possible values) of the goal attribute G occurring in the consequent. The log term is included in the formula (3) to normalize the value of RInt, so that this measure takes a value between 0 and 1. The InfoGain is given by:

$$InfoGain(A_i) = Info(G) - Info(G|A_i),$$
$$Info(G) = -\sum_{i=1}^{m_k}(p(g_i) \cdot \ln(p(g_i))),$$
$$Info(G|A_i) = \sum_{j=1}^{n}(p(v_{ij})(-\sum_{l=1}^{m_k} p(g_l|v_{ij}) \cdot \ln(p(g_l|v_{ij}))))$$

where m_k is the number of possible values of the goal attribute G_k, n_i is the number of possible values of the attribute A_i, $p(X)$ denotes the probability of X and $p(X|Y)$ denotes the conditional probability of X given Y.

Association Rule Mining: The fitness function in this case is also the same as the fitness function of classification rule mining with a little modification.

Confidence factor: The measure like confidence factor of association rule mining is the same as classification rule mining i.e.

$$C_f(R) = \frac{|A\&C|}{|A|}. \tag{1.6}$$

The only modification required is in comprehensibility and interestingness measures.

Comprehensibility: A careful study of the association rule infers that if the number of conditions involved in the antecedent part is less, the rule is more comprehensible. The following expression can be used to quantify the comprehensibility of an association rule.

$$C(R) = \frac{\ln(1 + |C|)}{\ln(1 + |A\&C|)}, \tag{1.7}$$

where $|C|$ and $|A\&C|$ are the number of attributes involved in the consequent part and the total rule, respectively.

Interestingness: As we mentioned earlier in the classification rules these measures can be defined by information theoretic measure (7). This way of measuring interestingness of the association rule will become computationally inefficient. For finding interestingness, the data set is to be divided based on each attribute present in the consequent part. Since a number of attributes can appear in the consequent part and they are not predefined, this approach may not be feasible for association rule mining. So a new expression is defined (which uses only the support count of the antecedent and the consequent parts of the rules) as

$$I(R) = (\frac{|A\&C|}{|A|}) \cdot (\frac{|A\&C|}{|C|}) \cdot (1 - \frac{|A\&C|}{|D|}), \tag{1.8}$$

where $|D|$ is the total number of records in the database.

Multiple Criteria in Data Clustering

This section provides an approach to data clustering based on the explicit optimization of a partitioning with respect to multiple complementary clustering objective (52). It has been shown that this approach may be more robust to the variety of cluster structures found in different data sets, and may be able to identify certain cluster structures that cannot be discovered by other methods. MOGA for data clustering uses two complementary objectives based on cluster compactness and connectedness. Let us define the objective functions separately.

Compactness: The cluster compactness can be measured by the overall deviation of a partitioning. This is simply computed as the overall summed distances between data items and their corresponding cluster center.

$$comp(S) = \sum_{c_k \in S} \sum_{i \in c_k} d(i, \mu_k), \tag{1.9}$$

where S is the set of all clusters, μ_k is the centroid of cluster c_k and $d(.)$ is the chosen distance function (e.g., Euclidean distance). As an objective, overall deviation should be minimized. This criterion is similar to the popular criterion of intra-cluster variance, which squares the distance value d(.) and is more strongly biased towards spherically shaped clusters.

Connectedness: This measure evaluates the degree to which neighboring data points have been placed in the same cluster. It is computed as

$$conn(S) = \sum_{i=1}^{N}(\sum_{j=1}^{L} x_{i,nn_i(j)}) \tag{1.10}$$

$nn_i(j)$ is the jth nearest neighbor of datum I and L is a parameter determining the number of neighbors that contributes to the connectivity measure. The connectivity should be minimized. After defining these two objectives, the algorithms given in Section 1.3, can be applied to optimize them simultaneously. The genetic operations such as crossover, mutation are similar to single objective genetic algorithm for data clustering. In addition, the following can also be used as criteria functions.

- The cohesiveness of clusters, which favors dense cluster.
- The distance between clusters and the global centroid, which favors well-separated clusters.
- The simplicity of the number of clusters, which favors candidate solutions with a smaller number of clusters.
- The simplicity of the selected attribute subset, which favors candidate solutions with a small number of selected attributes.

Each of the above measures is normalized into unit interval and is considered as an objective (to be maximized) in a multi-objective optimization problem.

1.3 Multi-objective Genetic Algorithms for KDD Issues

Since the KDD process involves many criteria like comprehensibility, predictive accuracy, and interestingness for classification rule mining (39; 40); it will be a suggestive approach to adapt multi-objective evolutionary algorithm for different types of multi-criteria optimization involved in our KDD process. A typical example is a scenario where one wants to maximize both the predictive accuracy and comprehensibility of an association rule. To optimize these three objectives simultaneously using evolutionary approach, this chapter provides a method called multi-objective genetic algorithm with a hybrid one-point uniform crossover. The pseudocode given here not only serves the task identified by us but also serves as a general framework for any kind of multi-criterion rule generation problem like association rule generation, fuzzy classification rule generation, dependency rule generation with an unknown parameter setting etc.

Pseudocode

1. $g = 1$; EXTERNAL(g)=ϕ;
2. Initialize Population $P(g)$;
3. Evaluate $P(g)$ by Objective Functions;
4. Assign Fitness to $P(g)$ Using Rank Based on Pareto Dominance
5. EXTERNAL(g)$\leftarrow$ Chromosomes Ranked as 1;
6. While (g $\leq$ Specified_no_of_Generation) do
7. $P'(g) \leftarrow$ Selection by Roulette Wheel Selection Schemes $P(g)$;
8. $P"(g) \leftarrow$ Single-Point Uniform Crossover and Mutation $P'(g)$;
9. $P"'(g) \leftarrow$ Insert/Remove Operation $P"(g)$;
10. $P(g+1) \leftarrow$ Replace $(P(g), P"'(g))$;
11. Evaluate $P(g+1)$ by Objective Functions;
12. Assign Fitness to $P(g+1)$ Using Rank Based Pareto Dominance;
13. EXTERNAL(g+1)$\leftarrow$[EXTERNAL(g)+Chromosome Ranked as One of P(g+1)];
14. $g = g + 1$;
15. End while
16. Decode the Chromosomes Stored in EXTERNAL as an IF-THEN Rule

1.3.1 Data Pre-processing

In general, the strength or quality of discovered knowledge entirely depends on the quality of the data being mined. This has motivated the KDD community to include data pre-processing as a preliminary steps. Though this step includes many tasks, in this chapter we will restrict ourselves to feature selection and instance selection only.

Feature Selection

The process of choosing a subset of features according to certain criteria, known as feature selection, has been extensively studied in the literature (9; 32). In particular, several different evolutionary algorithms (EA) have been employed as search strategies for feature selection (33; 34). In most cases, the aim is to optimize a single criterion, i.e. modeling accuracy, or a weighted-sum of accuracy and complexity. However, feature selection can naturally be posed as a multi-objective search problem, since in the simplest case it involves minimization of both the subset cardinality and modeling error. Therefore, multi-objective evolutionary algorithms (MOEA) are well suited for feature selection (46). In many problem domains, such as in medical or engineering diagnosis, performance maximization itself can comprise multiple objectives.

Instance Selection

Here, we describe two strategies for instance selection, which take part in most of the data mining algorithms.

Instance Selection for Prototype Selection: The 1-NN classifiers predict the class of a previously unseen instance by computing its similarity with a set of stored instances called prototypes. Storing a well-selected proper subset of available training instances has been shown to increase classifier accuracy in many domains. At the same time, the use of prototypes dramatically decreases both storage and classification-time costs. A prototype selection algorithm is an instance selection algorithm which attempts to obtain a subset of the training set that allows the 1-NN classifier to achieve the maximum classification rate. Each prototype selection algorithm is applied to an initial data set in order to obtain a subset of representative data items. We assess the accuracy of the subset selected using a 1-NN classifier.

Instance Selection for Training Set Selection: There may be situations in which a large number of data is present and this data, in most of the cases, is not equally useful in the training phase of a learning algorithm (19). Instance selection mechanisms have been proposed to choose the most suitable points in the data set to become instances for the training data set used by a learning algorithm. For example, in (19), a genetic algorithm (GA) is used for training data selection in radial basis function networks. Starting from the training data set the instance selection algorithm finds a suitable set, then a learning or DM algorithm is applied to evaluate each selected subset. This model is assessed using the test data set.

1.3.2 Data Mining Tasks

KDD refers to the overall process of turning low-level data into high-level knowledge. Data mining is one of the important steps of KDD process. A particular data mining algorithm is usually an instantiation of the model/preference/search components. The common algorithms in current data mining practice include the following:

1. Classification: classifies a data item into one of several predefined categorical classes.
2. Regression: maps a data item to a real-valued prediction variable.
3. Clustering: maps a data item into one of several clusters, where clusters are natural groupings of data items based on similarity matrices or probability density models.
4. Rule generation: extracts classification rules from the data.
5. Discovering association rules: describes association relationship among different attributes.
6. Summarization: provides a compact description for a subset of data.
7. Dependency modeling: describes significant dependencies among variables.
8. Sequence analysis: models sequential patterns like time-series analysis. The goal is to model the states of the process generating the sequence or to extract and report deviation and trends over time.

In this chapter, three important tasks of data mining, called classification, clustering and association rule generation are discussed.

Classification: This task has been studied for many decades by the machine learning and statistics communities (11; 35). In this task the goal is to predict the value (the class) of a user specified goal attribute based on the values of other attributes, called predicting attributes. For instance, the goal attribute might be the result of a B.Tech. student, taking the values (classes) "pass" or "fail", while the predicting attributes might be the student's age, branch, attendance, whether or not the student's grand total marks is above 60% etc.

Classification rules can be considered as a particular kind of prediction rules where the rule antecedent ("IF") part contains predicting attribute and rule consequent ("THEN") part contains a predicted value for the goal attribute. Examples of classification rules are:

IF(attendance > 75%)&(total_marks > 60%)THEN (result = "pass").
IF(attendance < 75%)THEN(result = "fail").

In the classification task the data being mined is divided into two mutually exclusive and exhaustive data sets: the training set and the test set. The data mining algorithm has to discover rules by accessing the training set only. In order to do this, the algorithm has access to the values of both the predicting attributes and goal attribute of each record in the training set. Once the training process is finished and the algorithm has found a set of classification rules, the predictive performance of these rules is evaluated on the test set, which was not seen during training. This is a crucial point. A measure of predictive accuracy is discussed in a later section. The application areas covered by classification task are credit approval, target marketing, medical diagnosis, treatment effectiveness etc. As already discussed in Section 2, the classification task is not only meant to discover highly predictive rule but also comprehensible and simultaneously interesting rules. Hence this can be viewed as a multi-objective problem rather than a single objective. The authors in (47) has given a multi-objective solution to this task by considering the aforesaid objectives (47).

Clustering: As mentioned above, in the classification task the class of a training example is given as input to the data mining algorithm, characterizing a form of supervised learning. In contrast, in the clustering task the data mining algorithm must, in some sense, "discover" classes by itself, by partitioning the examples into clusters, which is a form of unsupervised learning (12; 50). Examples that are similar to each other (i.e. examples with similar attribute values) tend to be assigned to the same cluster, whereas examples different from each other tend to be assigned to distinct clusters. Note that, once the clusters are found, each cluster can be considered as a class, so that now we can run a classification algorithm on the clustered data, by using the cluster name as a class label. GAs for clustering are discussed in (36; 37; 38). In cluster analysis of a particular data set, it is common to have a variety of cluster structures. Hence it is very difficult to detect clusters with different shapes and sizes by the algorithms considering only single criterion. To overcome the problems, many authors (52) have suggested different multi-objective clustering using evolutionary and non-evolutionary approaches.

Association Rule Mining: In the standard form of this task (ignoring variations proposed in the literature) each data instance (or "record") consists of a set of binary attributes called items. Each instance usually corresponds to a customer transaction, where a given item has a true or false value depending on whether or not the corresponding customer brought that item in that transaction. An association rule is a relationship of the form IF X THEN Y, where X and Y are sets of items and $X \cap Y = \phi$, (13; 48; 49; 53). An example of the association rule is:

IF fried_potatoes THEN soft_drink, ketchup.

Although both classification and association rules have an IF-THEN structure, there are important differences between them. We briefly mention here two of these differences. First, association rules can have more than one item in the rule consequent, whereas classification rules always have one attribute (the goal one) in the consequent. Second, unlike the association task, the classification task is asymmetric with respect to the predicting attributes and the goal attribute. Predicting attributes can occur only in the rule antecedent, whereas the goal attribute occurs only in the rule consequent. Like classification, association rule also has the objective to optimize the properties like comprehensibility, interestingness and confidence factor simultaneously. As it is already pointed out, the single criterion optimization algorithm is not up to the mark; and so the suggestive approach is to adapt a multi-criterion optimization algorithm.

1.3.3 Knowledge Post Processing

This includes some kind of post processing of the knowledge discovered by data mining algorithms. The motivation for such post processing stems from the large rule set generated by mining algorithms. To improve knowledge comprehensibility either we have to return back to any of the previous steps or we can undergo some kind of post processing techniques (e.g., removal of few rules/rule conditions). In addition, we often want to extract a subset of interesting rules among all discovered ones. The reason is that although many data mining algorithms were designed to discover accurate, comprehensible rules most of these algorithms were not designed to discover interesting rules.

1.4 Conclusions and Discussion

In this chapter, we have identified the multiple criteria implicitly/explicitly present in various steps of KDD. In addition, we have discussed genetic algorithms and their utility to multi-criteria problems like rule mining, data clustering and feature /instance selection. The use of multi-objective framework for the addressed tasks of KDD offers a great amount of flexibility that we hope to exploit in future work.

References

[1] Deb K (2001) Multi-objective optimisation using evolutionary algorithms. Wiley, New York
[2] Goldberg D E (1989) Genetic algorithms in search, optimisation and machine learning. Addison-Wesley
[3] Rechenberg I (1973) Evolutionsstrategie: optimierung technischer systeme, nach prinzipien der biologischen evolution. Frammann-Holzboog Verlag, Stuttgart
[4] Durham W (1994) Co-evolution:genes, culture, and human diversity. Stanford University Press
[5] Koza J R (1992) Genetic programming: on the programming of computers by means of natural selection. MIT Press, Cambridge, MA
[6] Back T (1996) Evolutionary algorithms in theory and practice. Oxford University Press, New York
[7] Freitas A A (2002) Data mining and knowledge discovery with evolutionary algorithms. Springer-Verlag, New York
[8] Cover J M, Thomas J A (1991) Elements of information theory. John Wiley
[9] Liu H, Motoda H (1998) Feature selection for knowledge discovery and data mining. Kluwer Academic Publishers
[10] Quinlan J R (1993) C4.5: programs for machine learning. Morgan Kaufmann, San Mateo, CA
[11] Fukunaga K (1972) Introduction to statistical pattern recognition. Academic Press, New York
[12] Jain A K, Dubes R C (1988) Algorithm for clustering data. Pretince Hall, Englewood Cliffs, New Jersey
[13] Adamo J M (2001) Data mining for association rules and sequential patterns. Springer-Verlag, New York
[14] Fogel D B (2000) Evolutionary computation: Toward a new philosophy of machine intelligence. IEEE Press, Piscataway, New Jersey
[15] Fogel D B (1998) Evolutionary computation: The fossil record. IEEE Press, Piscataway, New Jersey
[16] Fayyad U M, Piatetsky-Shapiro G, Smyth P (1996) From data mining to knowledge discovery: an Overview. In: Fayyad U M, Piatetsky-Shapiro G, Smyth P, Uthurusamy R (eds) Advances in Knowledge Discovery and Data Mining. MIT Press, MA 1–34
[17] Brachman R J, Anand T (1996) The process of knowledge discovery in databases: a human centred approach. In: Fayyad U M, Piatetsky-Shapiro G, Smyth P, Uthurusamy R (eds) Advances in Knowledge Discovery and Data Mining. MIT Press, MA 37–57
[18] Frawley W J, Piatetsky-Shapiro G, and Matheus C J (1991) Knowledge discovery in databases: an overview. In: Piatetsky-Shapiro G, Frawley B (eds) Knowledge Discovery in Databases. MIT Press, MA 1–27

[19] Reeves C R, Bush D R (2001) Using genetic algorithms for training data selection in RBF networks. In: Liu H, Motoda H (eds) Instance Selection and Construction for Data Mining, Norwell, MA: Kluwer 339–356
[20] Cano J R, Herrera F, Lozano M (2003) Using evolutionary algorithms as instance selection for data reduction in KDD: an experimental study. IEEE Transactions on Evolutionary Computation 7(6):561–575
[21] Ghosh A, Nath B (2004) Multi-objective rule mining using genetic algorithms. Information Sciences 163:123–133
[22] Babcock C (1994) Parallel processing mines retail data. Computer World, 6
[23] Fashman K H, Cuticchia A J, Kingsbury D T (1994) The GDB (TM) human genome database. Anna. Nucl. Acid. R. 22(17):3462–3469
[24] Weir N, Fayyad, U M, Djorgovski S G (1995) Automated star/galaxy classification for digitized POSS-II. Astron. Journal 109(6): 2401–2412
[25] Weir N, Djorgovski S G, Fayyad U M (1995) Initial galaxy counts from digitized POSS-II. Astron. Journal 110(1):1–20
[26] Back T, Schwefel H-P (1993) An overview of evolutionary algorithms for parameter optimisation. Evolutionary Computation 1(1):1–23
[27] Ghosh A, Dehuri S (2004) Evolutionary algorithms for multi-criterion optimization: a survey. International Journal on Computing and Information Science 2(1):38–57
[28] Laumanns M, Rudolph G, Schwefel H-P (1998) A spatial predator-prey approach to multi-objective optimization. Parallel Problem Solving from Nature 5:241–249
[29] Deb K, Pratap A, Agrawal S, Meyarivan T (2002) A Fast and elitist multi-objective genetic algorithm: NSGA-II. IEEE Transactions on Evolutionary Computation 6:182–197
[30] Ziztler E, Thiele L (1999) Multi-objective evolutionary algorithms: a comparative case study and strength pareto approach. IEEE Transactions on Evolutionary Computation 3:257–271
[31] Reinartz T (2002) A unifying view on instance selection. Data Mining and Knowledge Discovery 6:191–210
[32] Kohavi R, John G H (1997) Wrappers for feature subset selection. Artificial Intelligence 97:273–324
[33] Siedlecki W, Sklansky J (1989) A note on genetic algorithms for large scale feature selection. Pattern Recognition Letters 10:335–47
[34] Jain A, Zongker D (1997) Feature selection: evaluation, application and small sample performance. IEEE Transactions on Pattern Analysis and Machine Intelligence 19:153–158
[35] Lim T-S, Loh W-Y, Shih Y-S (2000) A Comparison of prediction accuracy complexity and training time of thirty-three old and new classification algorithms. Machine Learning Journal 40:203–228
[36] Krovi R (1991) Genetic algorithm for clustering: a preliminary investigation. IEEE Press 504–544
[37] Krishna K, Murty M N (1999) Genetic K-means algorithms. IEEE Transactions on Systems, Man and Cybernetics- Part-B 29:433–439

[38] Sarafis I, Zalzala AMS, Trinder P W (2001) A genetic rule based data clustering toolkit
[39] Dehuri S, Mall R (2006) Predictive and comprehensible rule discovery using a multi-objective genetic algorithm. Knowledge Based Systems 19(6):413–421
[40] Dehuri S, Patnaik S, Ghosh A, Mall R (2007) Application of elitist multi-objective genetic algorithm for classification rule generation. Applied Soft Computing (in press)
[41] Ishibuchi H, Nakashima T (2000) Multiobjective pattern and feature selection by a genetic algorithm. Proceedings of Genetic and Evolutionary Computation Conference 1069–1076
[42] Fayyad U M, Uthurusamy, R (1995) (eds) Proceedings 1st Internationl Conference Knowledge Discovery and Data Mining (KDD-95). AAAI Press, Montreal, Canada
[43] Simoudis E, Han J, Fayyad, U M (1996) Proceedings 2nd International Conference Knowledge Discovery & Data Mining, Portland, Oregon
[44] Holland J H (1990) ECHO: Explorations of evolution in a miniature world. In Famer J D, Doyne J (eds) Proceedings of the Second Conference on Artificial Life, Addison Wesley
[45] Noda E, Freitas A A, Lopes H S (1999) Discovering interesting prediction rules with a genetic algorithm. Proceeding Conference on Evolutionary Computation (CEC-99), Washington D.C., USA 1322–1329
[46] Emmanouilidis C, Hunter A, MacIntyre J (2000) A multiobjective evolutionary setting for feature selection and a commonality based crossover operator. Proceedings CEC-2000 La Jolla CA, USA 309–316
[47] Dehuri S, Mall R (2004) Mining predictive and comprehensible classification rules using multi-objective genetic algorithm. Proceeding of ADCOM, India
[48] Agrawal R, Srikant R (1994) Fast algorithms for mining association rules. Proceedings of the 20th International Conference on Very Large Databases
[49] Agrawal R, Imielinski T, Swami A (1993) Mining association rules between sets of items in large databases. Proceedings of ACM SIGMOD Conference on Management of Data 207–216
[50] Jie L, Xinbo G, Li-Chang J (2003) A GA based clustering algorithm for large datasets with mixed numeric and categorical values. Proceedings of the 5th International Conference on Computational Intelligence and Multi. Application
[51] Schewefel H -P (1975) Evolutionsstrategie and numerische optimierung. Ph. D. Thesis, Technical University, Berlin
[52] Handl J, Knowles J (2004) Multi-objective clustering with automatic determination of the number of clusters. Tech. Report TR-COMPSYSBIO-2004-02 UMIST, Manchester
[53] Ayad A M (2000) A new algorithm for incremental mining of constrained association rules. Master Thesis, Department of Computer Sciences and Automatic Control, Alexandria University
[54] Schaffer J D (1984) Multiple objective optimization with vector evaluated genetic algorithms. PhD Thesis, Vanderbilt University

2

Knowledge Incorporation in Multi-objective Evolutionary Algorithms

Ricardo Landa-Becerra, Luis V. Santana-Quintero, Carlos A. Coello Coello

CINVESTAV-IPN (Evolutionary Computation Group)
Departamento de Computación
Av. IPN No. 2508, Col. San Pedro Zacatenco
México, D.F. 07360, MEXICO
rlanda@computacion.cs.cinvestav.mx, lvspenny@hotmail.com
ccoello@cs.cinvestav.mx

Summary. This chapter presents a survey of techniques used to incorporate knowledge into evolutionary algorithms, with a particular emphasis on multi-objective optimization. We focus on two main groups of techniques: those that incorporate knowledge into the fitness evaluation, and those that incorporate knowledge in the initialization process and the operators of an evolutionary algorithm. Several techniques representative of each of these groups are briefly discussed, together with some examples found in the specialized literature. In the last part of the chapter, we provide some research ideas that are worth exploring in the future by researchers interested in this topic.

2.1 Introduction

In many disciplines, optimization problems have two or more objectives, which are normally in conflict with each other, and that we wish to optimize simultaneously. These problems are called "multi-objective", and their solution involves the design of different algorithms than when dealing with single-objective optimization problems. Multi-objective problems, in fact, normally give rise not to one, but to a set of solutions (called Pareto optimal set) which are all equally good.

Evolutionary algorithms have been very popular for solving multi-objective optimization problems (1; 2), mainly because of their ease of use, and their wide applicability. However, multi-objective evolutionary algorithms (MOEAs) tend to consume an important number of objective function evaluations, in order to achieve a reasonably good approximation of the Pareto front, even when dealing with benchmark problems of low dimensionality. This is a major concern when attempting to use MOEAs for real-world applications, since we normally can afford only a fairly limited number of fitness function evaluations in such cases.

R. Landa-Becerra et al.: *Knowledge Incorporation in Multi-objective Evolutionary Algorithms*, Studies in Computational Intelligence (SCI) **98**, 23–46 (2008)
www.springerlink.com

Despite these concerns, little efforts have been reported in the literature to reduce the computational cost of MOEAs, and several of them only focus on algorithmic complexity (see for example (13), in which little else can be done because of the theoretical bounds related to nondominance checking (21). It has been until relatively recently, that researchers have developed techniques to achieve a reduction of fitness function evaluations by exploiting knowledge acquired during the search (16). This knowledge can, for instance, be used to adapt the recombination and mutation operators in order to sample offspring in promising areas of the search space (as done through the use of cultural algorithms (8)). Knowledge of past evaluations can also be used to build an empirical model that approximates the fitness function to optimize. This approximation can then be used to predict promising new solutions at a smaller evaluation cost than that of the original problem (15; 16).

This chapter describes some of the possible schemes by which knowledge can be incorporated into an evolutionary algorithm, with a particular emphasis on MOEAs. The taxonomy of approaches that we will cover in this chapter is shown in Figure 2.1. In this proposed taxonomy, we divided the techniques for knowledge incorporation in two groups: (1) those that incorporate knowledge into the fitness evaluations, and (2) those which incorporate knowledge in the initialization process and the operators of an evolutionary algorithm. Each of these two groups has several ramifications (as shown in Figure 2.1), each of which are discussed in this chapter.

The remainder of this chapter is organized as follows. In Section 2.2, we discuss schemes that incorporate knowledge into the fitness evaluations of an evolutionary algorithm. The three schemes normally adopted (problem approximation, functional approximation, and evolutionary approximation) are all discussed in this section. Section 2.3 discusses the main schemes normally adopted to incorporate knowledge in the initialization and the operators of an evolutionary algorithm (namely, case-based reasoning and cultural algorithms). Finally, in Section 2.5, we provide some of the paths for future research within this topic that we consider worth exploring.

2.2 Knowledge Incorporation into the Fitness Evaluations

The high number of fitness evaluations often required by evolutionary algorithms is normally expensive, time-consuming or otherwise problematic in many real-world applications. Particularly in the following cases, a computationally efficient approximation of the original fitness function reducing either the number of duration of the fitness evaluations, is necessary:

- If the evaluation of the fitness function is computationally expensive.
- If the fitness function cannot be defined in an algebraic form (e.g., when the fitness function is generated by a simulator).
- If additional physical devices must be used to determine the fitness function and this requires human interaction.
- If parallelism is not allowed.
- If the total number of evaluations of the fitness function is limited by financial constraints.

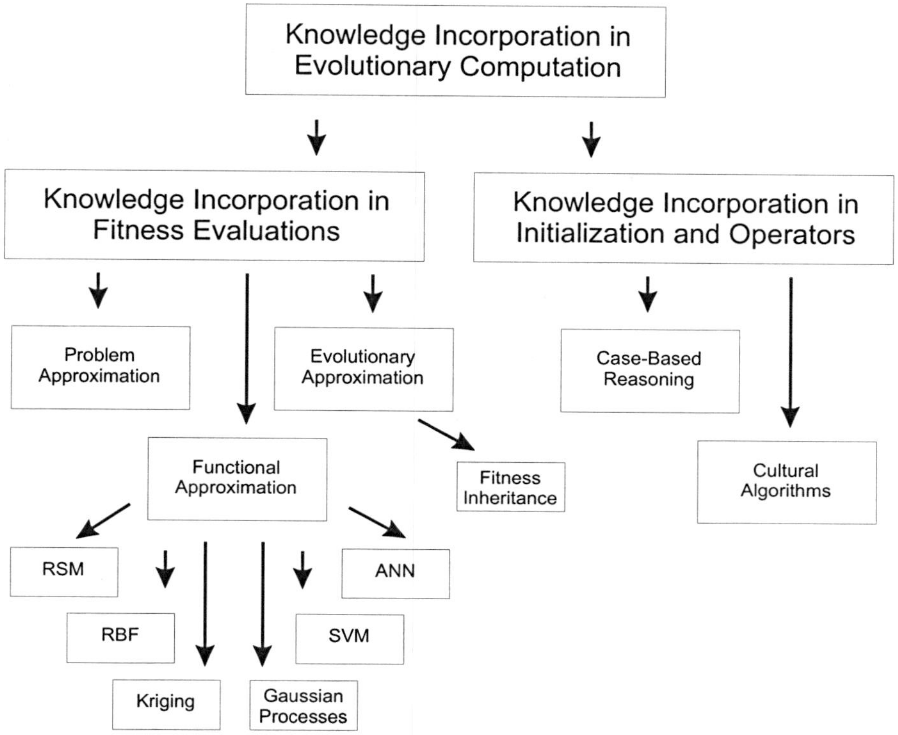

Fig. 2.1. A taxonomy of approaches for incorporating knowledge into evolutionary algorithms

In the above cases, the approximation is then used to predict promising new solutions at a smaller evaluation cost than that of the original problem. Jin (15) discusses various approximation levels or strategies adopted for fitness approximation:

Problem Approximation: Tries to replace the original statement of the problem by one which is approximately the same as the original problem but which is easier to solve. To save the cost of experiments, numerical simulations instead of physical experiments are used to pseudo-evaluate the performance of a design.

Functional Approximation: In this approximation, a new expression is constructed for the objective function based on previous data obtained from the real objective functions. In this case, models obtained from data are often known as meta-models or surrogates (see section 2.2.1)

Evolutionary Approximation: This approximation is specific for EAs and tries to save function evaluations by estimating an individual's fitness from other similar individuals. A popular subclass in this category is known as fitness inheritance (see section 2.2.1)

Currently, there exist several evolutionary algorithms that use a meta-model to approximate the real fitness function and reduce the total number of fitness evaluations without degrading the quality of the results obtained. To achieve this goal, meta-models should be combined with the original fitness function in a proper manner. The mechanism adopted to balance the use of the meta-model and the real objective function is known as *evolution control*. Evolution control takes an important role when meta-models are combined with the original fitness function. In such cases, most meta-models could converge to a local optimum if they are provided incorrect knowledge (or information) about the real function. There are two different forms to combine the approximated model and the real function:

Individual-based evolution control: In this case, some individuals use meta-models to evaluate their fitness value and others (in the same generation) use the real fitness function. The main issue in individual-based evolution control is to determine which individuals should use the meta-model and which ones should use the real fitness function for fitness evaluation. They can be randomly chosen from the population, or one could simply choose the best individuals in the population to be evaluated using the real function (see Figure 3.1).

Generation-based evolution control: The main issue in generation-based evolution control is to determine in which generations the meta-model should be used and in which generations the real fitness function should be used. In this control, the real fitness function is applied at every g generations, where g is predefined and fixed throughout the evolutionary process (see Figure 2.3).

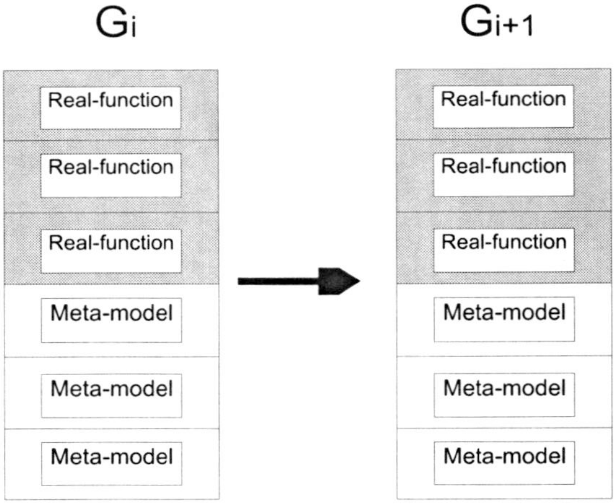

Fig. 2.2. Individual-based evolution control

In the above cases, the approximation is used to predict promising new solutions at a smaller evaluation cost than that of the original problem. Current functional approximation models include Polynomials (e.g., response surface methodologies (25; 74)), neural networks (e.g., multi-layer perceptrons (MLPs) (26; 27; 28)), radial-basis-function (RBF) networks (9; 63; 64), support vector machines (SVMs) (65;

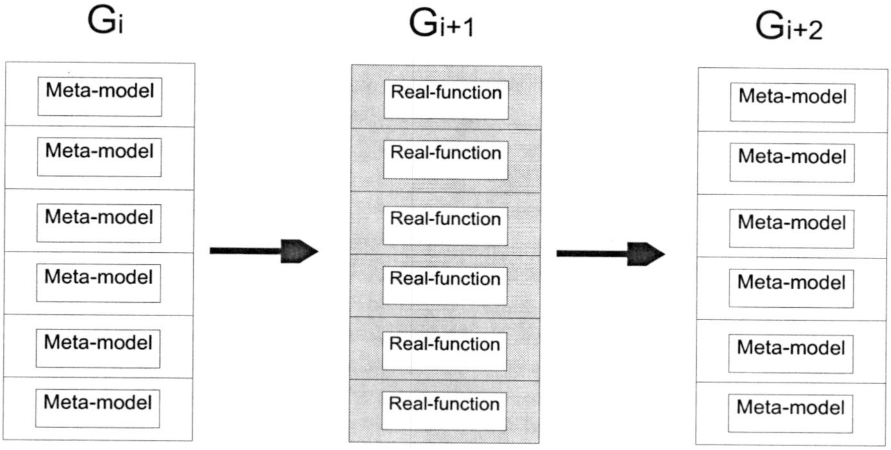

Fig. 2.3. Generation-based evolution control

66), Gaussian processes (29; 56), and Kriging (54; 55); all of them can be used for constructing meta-models.

Various approximation levels or strategies adopted for fitness approximation in evolutionary computation are proposed in (15).

2.2.1 Surrogates

Surrogate models can perform a number of tasks in support of a computational analysis. Through interpolation, extrapolation and/or integration, these models can be used to address complex problems involving experimental design, system analysis and prediction.

In a single-objective optimization context, surrogate models have been successful in dealing with highly demanding problems where the cost of evaluating the real fitness function is very expensive (computationally speaking).

The accuracy of the surrogate model relies on the number of samples provided in the search space, as well as on the selection of the appropriate model to represent the objective functions. There exist a variety of techniques for constructing surrogate models (see for example (3)). One approach is least-square regression using low-order polynomials, also known as response surface methods. A statistical alternative for constructing surrogate models is Kriging, which is also referred to as "Design and Analysis of Computer Experiments" (DACE) models (53) and Gaussian process regression (10). Comparisons of several surrogate modeling techniques are presented by Giunta and Watson (72) and by Jin et al. (73).

A surrogate model is built when the objective functions are to be estimated. This local model is built using a set of data points that lie on the local neighborhood of the design. Since surrogate models will probably be built thousands of times during the search, computational efficiency is the main objective. This motivates the use of radial basis functions, which can be applied to approximate multiple data, particularly when hundreds of data points are used for training.

Chafekar et al. (11) proposed a multi-objective evolutionary algorithm called OE-GADO, which runs several Genetic Algorithms (GAs) concurrently with each GA optimizing one objective function at a time, and forming a reduced model (based on quadratic approximation functions) with this information. At regular intervals, each GA exchanges its reduced model with the others. This technique can solve constrained optimization problems in 3,500 and 8,000 evaluations and is compared with respect to the NSGA-II (14) and the $\epsilon - MOEA$ (22; 60).

Emmerich et al. (23) present a local Gaussian random field meta-model (GRFM) to predict objective function values by exploiting information obtained during previous evaluations. This scheme was created for optimizing computationally expensive problems. This method selects the most promising population members at each generation so that they are evaluated using the real objective function. This approach was tested on a 10 dimensional airfoil optimization problem and was compared with respect to the NSGA-II in the generalized Schaffer problems.

Polynomials: Response Surface Methods (RSM)

The response surface methodology comprises regression surface fitting in order to obtain approximate responses, design of experiments in order to obtain minimum variances of the responses and optimizations using the approximated responses.

An advantage of this technique is that the fitness of the approximated response surfaces can be evaluated using powerful statistical tools. Additionally, the minimum variances of the response surfaces can be obtained using design of experiments with a small number of experiments.

For most response surfaces, the functions adopted for the approximations are polynomials because of their simplicity, although other types of functions are, of course, possible. For the cases of quadratic polynomials, the response surface is described as follows:

$$\hat{y} = (\beta_0) + \sum_{1 \leq i \leq k} (\beta_i * x_i) + \sum_{1 \leq i \leq j \leq n} (\beta_{k-1+i+j} * x_i * x_j) \qquad (2.1)$$

where k is the number of variables, and β_0 and β_i are the coefficients to be calculated. To estimate the unknown coefficients of the polynomial model, both the least squares method (LSM) and the gradient method can be used, but either of them requires at least the same number of samples of the real objective function than the k_t coefficients in order to obtain good results.

Goel et al. (74) is one of the few works that has used RSM in multi-objective problems. In this paper, the NSGA-II (14) an a local search strategy called "$\epsilon - constraint$" are adopted to generate a solution set that is used for approximating the Pareto optimal front by a response surface method (RSM). This method is applied to a rocket injector design problem.

There are few applications of the use of surrogates in evolutionary multi-objective optimization. Two of them are briefly discussed next.

Bramanti et al. (31) tried to reduce the computational cost of a multi-objective evolutionary algorithm using neural networks interpolation for building an objective response surface in order to find multiple trade-off solutions in electromagnetic design problems.

Wilson et al. (57) used two types of surrogate approximations (response surfaces and Kriging models) in order to reduce the computational expense of designing piezomorph actuators. The method shows that is flexible and can accommodate a wide variety of experimental designs and approximations. The authors also show that this method works well for both convex and non-convex Pareto fronts.

Radial Basis Functions

Radial Basis Functions (RBFs) were first introduced by R. Hardy in 1971 (30). This term is made up of two different words: *radial* and *basis functions*. A radial function refers to a function of the type:

$$g : \mathbb{R}^d \rightarrow \mathbb{R} : (x_1, \ldots, x_d) \mapsto \phi(\|x_1, \ldots, x_d\|_2),$$

for some function $\phi : \mathbb{R} \rightarrow \mathbb{R}$. This means that the function value of g at a point $\overrightarrow{x} = (x_1, \ldots, x_d)$ only depends on the Euclidean norm of $\overrightarrow{x}$:

$$\|\overrightarrow{x}\|_2 = \sqrt{\sum_{i=0}^{d} x_i^2} = \text{distance of } \overrightarrow{x} \text{ to the origin},$$

and this explains the term *radial*. The term *basis function* is explained next. Let's suppose we have certain points (called centers) $\overrightarrow{x}_1, \ldots, \overrightarrow{x}_n \in \mathbb{R}^d$. The linear combination of the function g centered at the points $\overrightarrow{x}$ is given by:

$$f : \mathbb{R}^d \mapsto \mathbb{R} : \overrightarrow{x} \mapsto \sum_{i=1}^{n} \lambda_i g(\overrightarrow{x} - \overrightarrow{x_i}) = \sum_{i=1}^{n} \lambda_i \phi(\|\overrightarrow{x} - \overrightarrow{x_i}\|) \qquad (2.2)$$

where $\|\overrightarrow{x} - \overrightarrow{x_i}\|$ is the Euclidean distance between the points $\overrightarrow{x}$ and $\overrightarrow{x}_i$. So, f becomes a function which is in the finite dimensional space spanned by the basis functions:

$$g_i : \overrightarrow{x} \mapsto g(\|\overrightarrow{x} - \overrightarrow{x_i}\|).$$

Now, let's suppose that we already know the values of a certain function $H : \mathbb{R}^d \mapsto \mathbb{R}$ at a set of fixed locations $\overrightarrow{x_i}, \ldots, \overrightarrow{x_n}$. These values are named $f_j = H(\overrightarrow{x_j})$, so we try to use the $\overrightarrow{x_j}$ as centers in the Equation 2.2. If we want to force the function f to take the values f_j at the different points $\overrightarrow{x_j}$, then we have to put some conditions on the λ_i. This implies the following:

$$\forall j \in \{1, \ldots, n\}\ f_j = f(\overrightarrow{x_j}) = \sum_{i=1}^{n} (\lambda_i \cdot \phi(\|\overrightarrow{x_j} - \overrightarrow{x_i}\|)) .$$

In these Equations, only the λ_i are unknown, and the Equations are linear in their unknowns. Therefore, we can write these Equations in matrix form:

$$\begin{bmatrix} \phi(0) & \phi(\|x_1 - x_2\|) & \dots & \phi(\|x_1 - x_n\|) \\ \phi(\|x_2 - x_1\|) & \phi(0) & \dots & \phi(\|x_2 - x_n\|) \\ \vdots & \vdots & & \vdots \\ \phi(\|x_n - x_1\|) & \phi(\|x_n - x_2\|) & \dots & \phi(0) \end{bmatrix} \cdot \begin{bmatrix} \lambda_1 \\ \lambda_2 \\ \vdots \\ \lambda_n \end{bmatrix} = \begin{bmatrix} f_1 \\ f_2 \\ \vdots \\ f_n \end{bmatrix}. \tag{2.3}$$

Typical choices for the kernel $g(\overrightarrow{x})$ include linear splines, cubic splines, multiquadrics, thin-plate splines and Gaussian functions as shown in Table 2.1.

Table 2.1. Radial basis functions

Type of Radial Function		
LS	linear splines	$\lvert r\rvert$
CS	cubic splines	$\lvert r\rvert^3$
MQS	multiquadrics splines	$\sqrt{1 + (\epsilon r)^2}$
TPS	thin plate splines	$\lvert r\rvert^{2m+1} \ln \lvert r\rvert$
GA	Gaussian	$e^{-(\epsilon r)^2}$

Ong et al. (32) used surrogate models (RBFs) to solve computationally expensive design problems with constraints. The authors used a combination of a parallel evolutionary algorithm coupled with sequential quadratic programming in order to find optimal solutions of an aircraft wing design problem. In this case, the authors construct a local surrogate model based on radial basis functions in order to approximate the objective and constraint functions of the problem.

Karakasis et al. (58) used surrogate models based on radial basis functions in order to deal with computationally expensive problems. A method called Inexact Pre-Evaluation (IPE) is applied into a MOEA's selection mechanism. Such method helps to choose the individuals that are to be evaluated using the real objective function, right after a meta-model approximation has been obtained by the surrogate. The results are compared against a conventional MOEA in two test-problems, one from a benchmark and one from the turbomachinery field.

Voutchkov & Keane (59) studied several surrogate models (RSM, RBF and Kriging) in the context of multi-objective optimization using the NSGA-II (14) as the MOEA that optimized the meta-model function given by the surrogate. The surrogate model is trained with 20 initial points and the NSGA-II is run on the surrogate model. Then, the 20 best resultant points given by the optimization are added to the existing data pool of real function evaluations and the surrogate is re-trained with these new solutions. A comparison of results is made in 4 test functions (from 2 to 10 variables), performing 400 real fitness function evaluations.

Kriging

In Kriging, the meta-model prediction is formed by adding up two different models as follows:

$$y(\overrightarrow{x}) = a(\overrightarrow{x}) + b(\overrightarrow{x})$$

where $a(\overrightarrow{x})$ represents the "average" long-term range behavior and the expected value of the true function. This function can be modeled in various ways, such as with polynomials or with trigonometric series as:

$$a(\overrightarrow{x}) = a_0 + \sum_{i=1}^{L} \sum_{j=1}^{R} a_{ij}(x_i)^j$$

where: R is the polynomial order with L dimensions and $b(\overrightarrow{x})$ stands for a local deviation term. $b(\overrightarrow{x})$ is a Gaussian random function with zero mean and non-zero covariance that represents a localized deviation from the global model. This function represents a short-distance influence of every data point over the global model. The general formulation for $b(\overrightarrow{x})$ is a weighted sum of N functions, $K_n(\mathbf{x})$ that represent the covariance functions between the nth data point and any point $\mathbf{x}$:

$$b(\overrightarrow{x}) = \sum_{n=1}^{N} b_n K(h(x, x_n)) \ \& \ h(x, x_n) = \sqrt{\sum_{i=1}^{L} (\frac{x_i - x_{in}}{x_i^{max} - x_i^{min}}^2)}$$

where x_i^{min} and x_i^{max} are the lower an upper bounds of the search space and x_{in} denotes the $i - th$ component of the data point x_n. However, the shape of $K(h)$ has a strong influence on the resulting aspect of the statistical model. And that is why it is said that Kriging is used as a estimator or an interpolator.

Knowles (16) proposed "ParEGO", which consists of a hybrid algorithm based on a single optimization model (EGO) and a Gaussian process, which is updated after every function evaluation, coupled to an evolutionary algorithm. EGO is a single-objective optimization algorithm that uses Kriging to model the search landscape from the solutions visited during the search and learns a model based on Gaussian processes (called DACE). This approach is used to solve multi-objective optimization problems of low dimensionality (up to 6 decision variables) with only 100 and 250 fitness function evaluations.

Neural Networks

Artificial neural networks (ANNs) provide a general, practical method for learning real-valued, discrete-valued, and vector-valued functions from examples. ANN learning is robust to errors in the training data and has been successfully applied to several problems. An ANN basically builds a map between a set of inputs and the respective outputs and are good to deal with nonlinear regression analysis and incomplete data. The main objective in the construction of an ANN is defining its appropriate architecture, that is, the number of layers and the number of nodes in each layer, given a certain number of inputs and outputs.

The Multi-Layer Perceptron (MLP) network has been widely used in function approximation problems, because it has been often found to provide compact representations of mappings in real-world problems. An MLP is composed of neurons which are very close to the ones represented in the case of the linear network. The linear neurons are modified so that a slight nonlinearity is added after the linear summation. The output (y) of each neuron is thus:

$$y = \phi(\sum_{i=1}^{n} w_i \cdot a_i + b)$$

where a_i are the inputs of the neuron and w_i are the weights of the neuron. The nonlinear function ϕ is called the activation function as it determines the activation level of the neuron.

In Figure 2.4, we show an MLP network with one layer of linear output neurons and one layer of nonlinear neurons between the input and output neurons. The middle layers are usually called *hidden layers*. Note that a graphical model of the conditional dependencies would not include the middle layer because the computational units are unknown variables of the model whereas the weights would be included as nodes of the model.

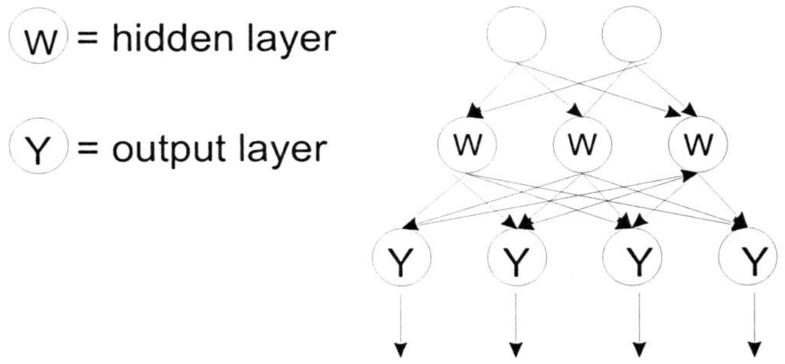

Fig. 2.4. A graphical representation of an MLP network with one hidden layer

Some applications of ANNs to evolutionary multi-objective optimization are the following:

Farina (35) proposed "NN-GRS", which is an extension of the single-objective Neural Network-based GRS (Generalized Response Surface) method. The main idea of this approach is to maintain two different objective functions, one real and another one which is an approximation (neural network based).

Nain & Deb (36) proposed "NSGA-II-ANN", which combines the NSGA-II algorithm (14) with a new method based on neural networks as the basic approximation technique for fitness computation. This meta-model is updated at each generation, and it provides a more refined approximate model to guide the search of the NSGA-II in subsequent generations.

Fitness Inheritance

Fitness Inheritance is a technique proposed by Smith et al. (38), whose main motivation is to reduce the total number of fitness evaluations made by an evolutionary algorithm. The mechanism works as follows: when assigning the fitness to an individual, sometimes the objective function is evaluated as usual, but the rest of the time, the fitness of an individual is assigned as the average of the finesses of its parents, thus avoiding a fitness function evaluation based on the assumption of similarity of the individual to its parents.

In the original proposal, inheritance was made in two ways: first, it is computed as the average of the two parents:

$$f(x) = \frac{f(p_1) + f(p_2)}{2}$$

where x is the individual to evaluate, and p_1 and p_2 are the parents of x. The second way is a weighted sum, where one of the parents may be more important. This weighted inheritance is obtained by:

$$f(x) = \frac{w_1 f(p_1) + w_2 f(p_2)}{w_1 + w_2}$$

where w_1 and w_2 are the weights for the two parents.

The OneMax problem was used in this early proposal (38), and the authors showed a decrease in the number of fitness function evaluations, when comparing this approach to another one without fitness inheritance. Other applications of fitness inheritance are: the design of vector quantization codebooks (61), where good results were obtained; and image compression (17), where the authors reported a significant reduction in the number of evaluations. This last approach is a more complex inheritance mechanism, where the individuals store also a variable that reflects the quality of the approximations with respect to the actual fitness values (a value of one means that the real objective function was evaluated). Such quality is decreased over generations, since the evolutionary algorithm tends to converge and the distance of an individual with respect to its parents tends to decrease over time.

Fitness inheritance must not be always applied, since the algorithm needs to use the true fitness function several times, in order to obtain enough information to guide the search. The percentage of time in which fitness inheritance is applied is called *inheritance proportion*. As this inheritance proportion tends to one (i.e., inherit all the time), the algorithm is most likely to prematurely converge (5).

There are a few theoretical studies about the fitness inheritance technique. For example, in (37), the authors present a theoretical model, which is used to obtain the convergence time, the optimal population sizing and the optimal inheritance proportion. Regarding the inheritance proportion, the authors found that, for problems of moderate and large size, the optimal values are between 0.54 and 0.558. They also defined the speedup of fitness inheritance when comparing it with respect to an algorithm without fitness inheritance (this is analogous to the speedup achieved when parallelizing an algorithm, with respect to executing in sequentially).

The work of Sastry et al. (37) was extended to the multi-objective case by Chen et al. (5). In this paper, the authors use fitness sharing to maintain diversity in the population an to be able to cover a larger extension of the Pareto front. The problem they solve is a bi-objective extension of the OneMax problem originally solved by Sastry et al. (37). Chen et al. (5) also present generalizations of the theoretical work reported in (37) about the convergence time, the optimal population sizing, the optimal inheritance proportion and the speedup. The experiments carried out show that, savings of up to 40% of the total number of evaluations can be achieved when using fitness inheritance alone. Additionally, when using fitness inheritance in combination with fitness sharing, savings of up to 25% can be obtained.

Bui el al. (39) presented a multi-objective approach for dealing with noisy functions. As they try to characterize the noise present, they evaluate a single individual several times, and this increases the computational cost associated with evaluations of the fitness function. They use fitness inheritance as an alternative to reduce the cost required by the approach. In their experiments they use the NSGA-II, with some specific anti-noise mechanisms, such as a probabilistic model and resampling. The test functions adopted are a noisy version of the ZDT set (12), which is a commonly used benchmark in evolutionary multi-objective optimization. The results show savings of up to 33% of the total number of fitness evaluations performed.

Fitness inheritance has also been used with particle swarm optimization. The main mechanism needs a modification in this case, because of the fact that in particle swarm optimization there are no parents or children, but leaders and previous positions of a particle. In (62), the authors propose several ways to make the inheritance, taking into account the previous position of the particle (the so called *pbest*), and the leader. They compare the performance of different versions of fitness inheritance in some multi-objective problems (the ZDT test functions (12) and the DTLZ test functions (6)), and also compare the approach with respect to other multi-objective versions of particle swarm optimization. The results show that the proposed approach is very competitive with respect to other techniques, providing savings that range between 30% and 40%

A more recent approach proposes the use of dynamic rules to assign the inheritance proportion in particle swarm optimization (40). This scheme is proposed again for multi-objective problems. Five rules were proposed in (40):

$$p_i(gen) = \left(\frac{gen}{Gmax}\right)^4$$

$$p_i(gen) = \left(\frac{gen}{Gmax}\right)^2$$

$$p_i(gen) = \frac{gen}{Gmax} - \frac{\sin\left(2\pi\frac{gen}{Gmax}\right)}{6.3}$$

$$p_i(gen) = \frac{gen}{Gmax}$$

$$p_i(gen) = \left(\frac{gen}{Gmax}\right)^{1/2}$$

$$p_i(gen) = \left(\frac{gen}{Gmax}\right)^{1/4}$$

where $p_i(gen)$ is the inheritance proportion at generation gen, and $Gmax$ is the total number of generations that the algorithm will run.

The use of these rules results on savings from 19% up to 78% of the total number of evaluations. However, the greater the savings in the number of evaluations, the greater is the degradation in the quality of the results. Nevertheless, the authors show that functions that provide up to 49% of savings present no significant loss in the quality of the results. The approach was tested with the well known ZDT problems (12).

As a final note on fitness inheritance, it is important to mention that some researchers consider this mechanism not so useful in complex or real world problems, because it has been only applied to "easy" problems. For example (67) tested the original scheme of fitness inheritance on a standard binary GA and the ZDT problems (12). From this study, the authors concluded that fitness inheritance is not useful when the shape of the Pareto front is nonconvex or discontinuous. These conclusions are true for the original proposal, but it is possible to overcome them, as shown in (71), so that fitness inheritance can be successfully adopted with Pareto fronts of any shape.

2.3 Knowledge Incorporation in the Initialization and the Operators

In addition to the techniques for incorporating knowledge during the evaluation of new individuals (i.e., directly affecting fitness calculation), some researchers have explored the use of knowledge at different stages of the evolutionary search. The aim, however, is the same: either to speedup convergence or to reduce the computational cost of the algorithm (measured in terms of the number of fitness function evaluations performed).

In this regard, several options are possible. For example, some domain knowledge previously obtained (or provided by a human expert) may be used to build a database of cases that are adopted as a reference for conducting a more efficient search. It is also possible to seed good solutions in the initial population of an evolutionary algorithm, or to add knowledge (perhaps based on experience) to the variation operators (recombination and mutation), in order to facilitate the generation of good solutions. Next, we will these choices in more detail.

2.3.1 Case-Based Reasoning

The case-based reasoning method (4) consists of a historical database with cases of study of the problem to be solved. When new problems appear, the system looks for a case in the database which matches the current problem, and adapts the previous solutions. Sometimes the system can even store the process to obtain a solution, in order to execute the same or a similar process when similar problems appear.

Early applications of case-based reasoning in evolutionary computation consist only of a tool to analyze the evolutionary process, making use of the building blocks theory (19). The authors in this case, used several test functions, including the circuits design problem, and they applied the case-based reasoning techniques to discover the search path the algorithm followed, basically through the discovery of "good" building blocks. The authors argue that these tools can be used to post-process solutions, and they are also useful when dealing with deceptive problems.

An attempt to improve the performance of an evolutionary algorithm with case-based reasoning, is the use of a memory that contains previous solutions, obtained from similar problems (41). The population is initialized in this case with some of these solutions, in order to move the search to potentially good regions. Louis (41) proposed a methodology for selecting appropriate initialization cases. The approach works very well in combinatorial circuit design problems, showing that the time required to solve a problem decreases with respect to that required by the original algorithm. A similar application of case-based reasoning techniques to the extraction of design patterns within the context of combinational circuit design is reported by Islas et al. (18).

Similar approaches of injection of solutions obtained from a memory of previous runs were used to solve different instances of the traveling salesman problem (42), for dynamic strike force asset allocation (43), for electronic circuits design (43), and for games (44).

In the case of (42), the strategy is not only to inject potential solution at the beginning of the evolutionary process, but also periodically during the search. The authors also develop a similarity measure to select the appropriate cases for the traveling salesman problem.

In (43), the authors investigate the number of individuals that can be injected to the algorithm while it obtains good results. They conclude that if too many individuals are injected, the algorithm is more likely to have premature convergence (a similar feature present when using surrogates and fitness inheritance), and recommend to inject only between 5% and 15% of solutions.

More recent work investigates the changes in the convergence rates when applying injection of solutions (45), the identification of the similarity of the solved problems, and those to be solved.

Despite the lack of applications of case-based reasoning in evolutionary multi-objective optimization, there are a very interesting potential applications of this technique, including a multiple objective version of the traveling salesman problem, and several types of scheduling problems. These and many other applications could certainly benefit from the use of databases of cases and/or from domain knowledge provided by human experts.

2.4 Cultural Algorithms

Cultural algorithms were proposed by Robert Reynolds (7), as an approach that tries to add domain knowledge to an evolutionary algorithm during the search process, avoiding the need to add it *a priori*. This approach uses, in addition to the population space commonly adopted in evolutionary algorithms, a belief space, which encodes the knowledge obtained from the search points and their evaluation, in order to influence the evolutionary operators that guide the search. However, the belief space is commonly designed based on the group of problems that is to be solved.

At each generation, the cultural algorithm selects some exemplar individuals from the population, in order to extract information from them that can be useful during the search. Such an information is used to update the belief space. The belief space will then influence the operators of the evolutionary algorithm, to transform them in informed operators and enhance the search process. These interactions between the spaces of a cultural algorithm are depicted in Figure 2.5.

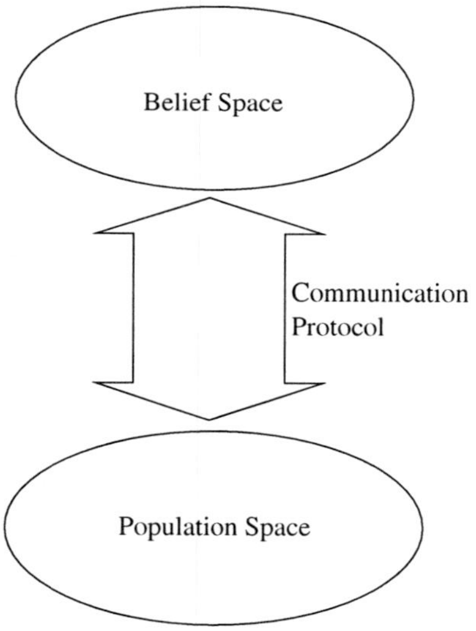

Fig. 2.5. Spaces of a cultural algorithm

Regarding optimization, cultural algorithms have been applied mainly to single-objective problems. One of the first applications was the Boole problem (7), which uses a genetic algorithm for the population space, and version spaces for the belief space. In this case, a graph (the belief space) of the schemes in the population is built and classified based on the fitness of each particular instance. This is a very illustrative approach, because the graph's dynamics will reflect the discovering of good and bad schemes.

Constrained and dynamic optimization problems have been a a very active application area for cultural algorithms. Reynolds et al. (8) proposed a cultural version of GENOCOP for solving convex constrained optimization problems. The belief space in this approach constructs boundaries to include the feasible region, and then produce new solutions only within that region.

Later on, Chung and Reynolds (68) developed the CAEP (Cultural Algorithms with Evolutionary Programming) for global optimization, with a very rich model in the belief space, and very encouraging results. For example, in (70) the authors proposed a formal model of self-adaptation in cultural algorithms, that supports the three main levels of self-adaptation in evolutionary algorithms (population, individual and component level). The royal road functions were adopted as a case of study in this work.

The CAEP was tested on a number of global optimization problems (20), showing its improvement when compared to the standard evolutionary programming algorithm. In this work, the belief space was divided in two parts, called knowledge sources, specifically designed for real-valued problems: the situational knowledge and the normative knowledge. A general description of these knowledge sources, and those designed and added later, is provided in Table 2.2.

Table 2.2. Knowledge sources for real-parameter optimization in cultural algorithms

Knowledge source	**Description**
Situational knowledge	Consists of the best exemplars found in the population, which represent leaders to follow. The individuals generated through this source, will tend to be closer to the leaders.
Normative knowledge	Consists of a set of intervals for each decision variable where good solutions have been found. The individuals generated through this source are more likely to be within the intervals, so they exploit good regions.
Topographical knowledge	Consists of a set of cells that represent a region of the search space. Each cell stores a characteristic of the region it represents; for example, the feasibility of that region. The individuals generated through this source will be closer to the best cells.
History knowledge	Consists of a set of previous local optima, and its function is to extract patterns about their position. The individuals generated through this source will try to find in advance the location of the next local optimum. This knowledge source can also be used to add diversity to the algorithm, since it attempts to explore new regions.
Domain knowledge	It has no defined structure, because it depends of the problem which is to be solved. Its function is to exploit some knowledge about the problem, if available.

A CAEP for nonlinear constrained optimization, a more general problem than the one tackled in GENOCOP, was proposed in (46). To handle constraints, a third knowledge source, called topographical knowledge, was added to the belief space. It consists of a set of cells, which store some characteristic of the region of the search space they represent. In this case, they store the feasibility of the region, based on the explored points within them.

Jin's approach (46) was extended in (47), improving its computational efficiency, and overcoming its scalability problems. In the original approach the topographical knowledge was stored as a n-dimensional grid of the search space. This was replaced by a spatial data structure, that requires a controlled amount of memory even when the number of dimensions grows. Additionally, the authors present an empirical study in which this approach is validated using a well-known benchmark adopted in evolutionary constrained optimization, and results are compared with respect to constraint-handling techniques representative of the state-of-the-art in the area. The cultural algorithm reported in (47) is able to find competitive results, while performing only about 15% of the total number of fitness function evaluations required by the other approaches with respect to which it was compared.

Saleem and Reynolds (69) added two more knowledge sources to cultural algorithms, in order to deal with dynamic environments: history knowledge and domain knowledge. The first of these sources was designed to extract patterns about the changes of position of optimal points at each environmental change. The second source was designed to exploit the known characteristics of the function generator. Even when these knowledge sources were designed for dynamic problems, they have also been used in static environments (48; 49).

Another cultural algorithm in which the population space adopts particle swarm optimization was proposed by Peng et al. (48). The authors use all of the previously designed knowledge sources, and they investigate the role of the belief space in the different stages of the optimization process.

Differential evolution has also been used for the population space of a cultural algorithm (33; 49), also focusing on constrained optimization. In this case, all the knowledge sources were adapted for its use with the differential evolution operator, providing some adaptation of its components. The approach was tested on a well known benchmark and also on some engineering optimization problems, with good results.

Finally, there is a cultural algorithm in which genetic programming is used for the population space (9). In this case, the belief space consists of a set of subgraphs which have been frequently found in the population.

There are very few attempts to use cultural algorithms for multi-objective optimization. The first is a CAEP (51), which uses Pareto ranking, and an approximation of the dimensions of the Pareto front in the belief space. The belief space works as a guide for the individuals to reach regions where nondominated solutions have been found. The belief space includes also a mechanism to obtain a good distribution of the resulting points along the Pareto front.

Recently, the cultural algorithm with differential evolution ((49)) for constrained optimization was proposed, together with the ε-constraint method, to solve multi-objective problems (50). In this work, the authors identify some hard problems from the DTLZ (6) and WFG (34; 52) benchmarks, and then they apply their technique to them. This approach is computationally expensive (because of the optimization processes to obtain one point at a time), but this cost, when affordable, is worth trying, because this approach can solve very difficult problems that other modern multi-objective evolutionary algorithms (e.g., the NSGA-II (14)) cannot solve, even if allowed to perform a very high number of fitness function evaluations.

2.5 Future Perspectives

From our previous discussion of the most representative work regarding incorporation of knowledge into evolutionary algorithms, we could identify several topics in which researchers working on evolutionary multi-objective optimization have not done much work (or where there is no work at all). Some of these topics are the following:

- One of the key issues when adopting meta-models for reducing the number of evaluations of a MOEA, is the design of the *evolution control* that keeps a proper balance between the evaluations done in the meta-model and those performed with the actual objective function. More detailed studies regarding the design of such evolution control are necessary in the context of multi-objective optimization.
- To the authors' best knowledge, so far, nobody has used support vector machines within the context of evolutionary multi-objective optimization (i.e., for predicting good solutions).
- The development of cultural algorithms for multi-objective optimization is another promising path for future research. The aim in this case, should be not only to reduce the number of fitness function evaluations to be performed, but also to be able to solve problems that are very hard for other multi-objective evolutionary algorithms.
- Case-based reasoning techniques have not been used so far (to the authors' best knowledge) within the context of evolutionary multi-objective optimization. This sort of approach may be very useful to seed good solutions in the initial population of a MOEA, or to extract certain "patterns" from the population of a MOEA that could be used to speedup convergence in similar multi-objective optimization problems.
- Scalability is another important topic that needs to be studied in more depth. In the few papers that we reviewed in this chapter, MOEAs that benefit from knowledge incorporation (particularly those adopting meta-models) have been used only in problems with few decision variables (no more than ten). Thus, the scalability of current techniques to large dimensional problems is a very important research topic that certainly deserves attention.

Acknowledgments

The first and second authors acknowledge support from CONACyT through a scholarship to pursue graduate studies at the Computer Science Department of

CINVESTAV-IPN. The third author acknowledges support from CONACyT project number 42435-Y.

References

[1] Coello Coello C A, Van Veldhuizen D A, Lamont G B (2002) Evolutionary algorithms for solving multi-objective problems. Kluwer Academic Publishers, New York

[2] Deb K (2001) Multi-objective optimization using evolutionary algorithms. John Wiley & Sons, Chichester, UK

[3] Vapnik V (1998) Statistical learning theory. Wiley Information Technology Encyclopedia and Acronyms. Springer, Berlin Heidelberg, New York

[4] Riesbeck C K, Schank R C (1989) Inside case-based reasoning. Lawrence Erlbaum Associates, New Jersey

[5] Chen J H, Goldberg D E, Ho S-Y, Sastry K (2002) Fitness inheritance in multi-objective optimization. In: Langdon W B and Cantú-Paz E, Mathias K, Roy R, Davis D, Poli R, Balakrishnan K, Honavar V, Rudolph G, Wegener J, Bull L, Potter M A, Schultz A C, Miller J F, Burke E, Jonoska N (eds) Proceedings of the Genetic and Evolutionary Computation Conference (GECCO'2002). Morgan Kaufmann Publishers, San Francisco, California, 319–326

[6] Deb K, Thiele L, Laumanns M, Zitzler E (2005) Scalable test problems for evolutionary multiobjective optimization. In: Abraham A, Jain L, Goldberg R (eds) Evolutionary Multiobjective Optimization. Theoretical Advances and Applications. Springer, USA, 105–145

[7] Reynolds R G (1994) An introduction to cultural algorithms. In: Sebald A V, Fogel L J (eds) Proceedings of the Third Annual Conference on Evolutionary Programming. World Scientific, 131–139

[8] Reynolds R G, Michalewicz Z, Cavaretta M (1995) Proceedings of the Fourth Annual Conference on Evolutionary Programming. In: McDonnell J R, Reynolds R G, Fogel, D B (eds) Using Cultural Algorithms for Constraint Handling in GENOCOP. MIT Press, 298–305

[9] Ong Y S, Nair P B, Keane A J, Wong K W (2004) Surrogate-assisted evolutionary optimization frameworks for high-fidelity engineering design problems. In: Jin Y (eds) Knowledge Incorporation in Evolutionary Computation, Springer, Studies in Fuzziness and Soft Computing, 307–332

[10] Williams C K I, Rasmussen C E (1996) Gaussian processes for regression. In: Touretzky D S, Mozer M C, Hasselmo M E (eds) Advances in Neural Information Processing Systems 8, MIT Press

[11] Chafekar D, Shi L, Rasheed K, Xuan J (2005) Multi-objective GA optimization using reduced models. IEEE Transactions on Systems, Man, and Cybernetics: Part-C 35 (2):261–265

[12] Zitzler E, Deb K, Thiele L (2000) Comparison of multiobjective evolutionary algorithms: empirical results. Evolutionary Computation 8(2):173–195

[13] Jensen M T (2003) Reducing the run-time complexity of multiobjective EAs: the NSGA-II and other algorithms. IEEE Transactions on Evolutionary Computation 7(5):503–515
[14] Deb K, Pratap A, Agarwal S, Meyarivan T (2002) A fast and elitist multiobjective genetic algorithm: NSGA–II. IEEE Transactions on Evolutionary Computation 6(2):182–197
[15] Jin Y (2005) A comprehensive survey of fitness approximation in evolutionary computation. Soft Computing 9(1):3-12
[16] Knowles J (2006) ParEGO: A hybrid algorithm with on-line landscape approximation for expensive multiobjective optimization problems. IEEE Transactions on Evolutionary Computation 10(1): 50–66
[17] Salami M, Hendtlass T (2003) A fast evaluation strategy for evolutionary algorithms. Applied Soft Computing 2:156–173
[18] Islas Pérez E and Coello Coello C A, Hernández Aguirre A (2005) Extraction and reuse of design patterns from genetic algorithms using case-based reasoning. Soft Computing–A Fusion of Foundations, Methodologies and Applications 9(1):44–53
[19] Louis S J, McGraw G, Wyckoff R (1993) Case-based reasoning assisted explanation of genetic algorithm results. Journal of Experimental and Theoretical Artificial Intelligence 5: 21–37
[20] Chung C-J, Reynolds R G (1998) CAEP: An evolution-based tool for real-valued function optimization using cultural algorithms. Journal on Artificial Intelligence Tools 7(3): 239–292
[21] Kung H T, Luccio F, Preparata F P (1975) On finding the maxima of a set of vectors. Journal of the Association for Computing Machinery 22(4): 469–476
[22] Deb K, Mohan M, Mishra S (2005) Evaluating the ϵ-domination based multi-objective evolutionary algorithm for a quick computation of pareto-optimal solutions. Evolutionary Computation 13(4):501–525
[23] Emmerich M T M, Giannakoglou K C, Naujoks B (2006) Single and multiobjective evolutionary optimization assisted by Gaussian random field metamodels. IEEE Transactions on Evolutionary Computation 10(4):421-439
[24] Gibbs M N, MacKay D J C (1996) Efficient implementation of Gaussian processes for interpolation. http://www.inference.phy.cam.ac.uk/mackay/abstracts/gpros.html
[25] Rasheed K, Ni X, Vattam S (2003) Comparison of methods for developing dynamic reduced models for design optimization. Soft Computing Journal (in press)
[26] Hong Y-S, Lee H, Tahk M-J (2003) Acceleration of the convergence speed of evolutionary algorithms using multi-layer neural networks. Engineering Optimization 35(1): 91–102
[27] Hüscken M, Jin Y, Sendhoff B (2005) Structure optimization of neural networks for aerodynamic optimization. Soft Computing Journal 9(1):21–28
[28] Pierret S (1999) Turbomachinery blade design using a navier-stokes solver and artificial neural network. ASME Journal of Turbomachinery 121(3): 326–332

[29] Bueche D, Schraudolph N N, Koumoutsakos P (2004) Accelerating evolutionary algorithms with Gaussian process fitness function models. IEEE Trans. on Systems, Man, and Cybernetics: Part C
[30] Hardy R L (1971) Multiquadric equations of topography and other irregular surfaces. Journal of Geophysics Research 76:1905–1915
[31] Bramanti A, Barba P Di, Farina M, Savini A (2001) Combining response surfaces and evolutionary strategies for multiobjective Pareto-optimization in electromagnetics. International Journal of Applied Electromagnetics and Mechanics 15(1):231–236
[32] Ong Y S, Nair P B, Keane A J (2003) Evolutionary optimization of computationally expensive problems via surrogate modeling. AIAA Journal 41(4):687–696
[33] Becerra R L, Coello Coello C A (2006) Cultured differential evolution for constrained optimization. Computer Methods in Applied Mechanics and Engineering 195 (33-36):4303–4322
[34] Huband S, Hingston P, Barone L, While L (2006) A review of multiobjective test problems and a scalable test problem toolkit. IEEE Transactions on Evolutionary Computation, 10(5): 477–506
[35] Farina M (2002) A neural network based generalized response surface multiobjective evolutionary algorithms. Congress on Evolutionary Computation, IEEE Press, 956–961
[36] Nain P K S, Deb K (2003) Computationally effective search and optimization procedure using coarse to fine approximation. Congress on Evolutionary Computation, IEEE Press, 2081–2088
[37] Sastry K, Goldberg D E, Pelikan M (2001) Don't evaluate, inherit. Proceedings of Genetic and Evolutionary Computation Conference, Morgan Kaufmann Publishers, 551–558
[38] Smith R E, Dike B A, Stegmann S A (1995) Fitness inheritance in genetic algorithms. SAC '95: Proceedings of the 1995 ACM symposium on Applied Computing, Nashville, Tennessee, United States, ACM Press, New York, NY, USA, 345–350
[39] Bui L T, Abbass H, Essam D (2005) Fitness inheritance for noisy evolutionary multi-objective optimization. Proceedings of Genetic and Evolutionary Computation Conference (GECCO-2005), ACM, 779–785
[40] Reyes-Sierra M, Coello Coello C A (2006) Dynamic fitness inheritance proportion for multi-objective particle swarm optimization. Genetic and Evolutionary Computation Conference (GECCO'2006), ACM Press. ISBN 1-59593-186-4, Seattle, Washington, USA, 89–90
[41] Louis S J, Johnson J (1997) Solving similar problems using genetic algorithms and case-based memory. Proceedings of the Seventh International Conference on Genetic Algorithms (ICGA97), Morgan Kaufmann, San Francisco, CA.
[42] Louis S J, Zhang Y (1999) A sequential similarity metric for case injected genetic algorithms applied to TSPs. Proceedings of the Genetic and Evolutionary Computation Conference, Morgan Kaufman, Orlando, Florida, USA, 377–384

[43] Louis S J (2002) Genetic learning for combinational logic design. Proceedings of the GECCO-2002 Workshop on Approximation and Learning in Evolutionary Computation, New York, NY, 21–26
[44] Miles C, Louis S J (2005) Case-injection improves response time for a real-time strategy game. Proceedings of the IEEE Symposium on Computational Intelligence in Games, IEEE Press, New York, NY
[45] Drewes R, Louis S J, Miles C, McDonnell J, Gizzi N (2003) Use of case injection to bias genetic algorithm solution of similar problems. Proceedings of the International Congress on Evolutionary Computation, IEEE Press, Canberra, Australia
[46] Jin X, Reynolds R G (1999) Using knowledge-based evolutionary computation to solve nonlinear constraint optimization problems: a cultural algorithm approach. Congress on Evolutionary Computation, IEEE Service Center, 1672–1678
[47] Coello Coello C A, Landa Becerra R (2002) Adding knowledge and efficient data structures to evolutionary programming: A cultural algorithm for constrained optimization. Proceedings of the Genetic and Evolutionary Computation Conference (GECCO'2002), Morgan Kaufmann Publishers, 201–209
[48] Peng B, Reynolds R G, Brewster J (2003) Cultural swarms. Proceedings of the Congress on Evolutionary Computation 2003 (CEC'2003), IEEE Service Center
[49] Landa Becerra R, Coello Coello C A (2005) Optimization with constraints using a cultured differential evolution approach. Genetic and Evolutionary Computation Conference (GECCO'2005), ACM Press,Washington, DC, USA, 1: 27–34
[50] Landa Becerra R, Coello Coello C A (2006) Solving hard multiobjective optimization problems using e-Constraint with cultured differential evolution. Parallel Problem Solving from Nature - PPSN VIII, LNCS, Springer-Verlag
[51] Coello Coello C A, Landa Becerra R (2003) Evolutionary multiobjective optimization using a cultural algorithm. IEEE Swarm Intelligence Symposium Proceedings, IEEE Service Center, 6–13
[52] Huband S, Barone L, While L, Hingston P (2005) A scalable multi-objective test problem toolkit. Evolutionary Multi-Criterion Optimization. Third International Conference, EMO 2005, Springer. Lecture Notes in Computer Science, 3410:280–295
[53] Sacks J, Welch W, Mitchell T, Wynn H (1989) Design and analysis of computer experiments (with discussion). Statistical Science 4:409–435
[54] Emmerich M, Giotis A, Özdenir M, Bäck T, Giannakoglou K (2002) Metamodel-assisted evolution strategies. Parallel Problem Solving from Nature, Lecture Notes in Computer Science, Springer, 371–380
[55] Ratle A (1998) Accelerating the convergence of evolutionary algorithms by fitness landscape approximation. Parallel Problem Solving from Nature V, 87–96

[56] Ulmer H, Streichert F, Zell A (2003) Evolution startegies assisted by Gaussian processes with improved pre-selection criterion. Proceedings of IEEE Congress on Evolutionary Computation, 692–699
[57] Wilson B, Cappelleri D J, Simpson T W, Frecker M I (2000) Efficient pareto frontier exploration using surrogate approximations. Symposium on Multidisciplinary Analysis and Optimization, AIAA, Long Beach, CA
[58] Karakasis M K, Giannakoglou K C (2005) Metamodel-assisted multi-objective evolutionary optimization. EUROGEN 2005. Evolutionary Methods for Design, Optimization and Control with Applications to Industrial Problems, Munich, Germany
[59] Voutchkov I, Keane A J (2006) Multiobjective optimization using surrogates. Adaptive Computing in Design and Manufacture Proceedings of the Seventh International Conference, the Institute for People-centered Computation (IP-CC), Bristol, UK, 167–175
[60] Deb K, Mohan M, Mishra S (2003) Towards a quick computation of well-spread pareto-optimal solutions. Evolutionary Multi-Criterion Optimization. Second International Conference, EMO 2003, Springer. Lecture Notes in Computer Science. 2632:222–236
[61] X. Zheng and B. A. Julstrom and W. Cheng (1997) Design of vector quantization codebooks using an genetic algorithm. Proceedings of IEEE International Conference on Evolutionary Computation (ICEC 97), 525–530
[62] Reyes Sierra M, Coello Coello C A (2005) A study of fitness inheritance and approximation techniques for multi-objective particle swarm optimization. IEEE Congress on Evolutionary Computation (CEC'2005), IEEE Service Center, Edinburgh, Scotland
[63] Ulmer H, Streicher F, Zell A (2003) Model-assisted steady-state evolution strategies. Proceedings of Genetic and Evolutionary Computation Conference, LNCS 2723:610–621
[64] Won K S, Ray T (2004) Performance of kriging and cokriging based surrogate models within the unified framework for surrogate assisted optimization. IEEE Congress on Evolutionary Computation, 1577–1585
[65] Abboud K, Schoenauer M (2002) Surrogate deterministic mutation. Proceedings of Artificial Evolution, 103–115
[66] Bhattacharya M, Lu G (2003) A dynamic approximate fitness based hybrid EA for optimization problems. Proceedings of IEEE Congress on Evolutionary Computation, 1879–1886
[67] Ducheyne Els I, De Baets B, De Wulf R (2003) Is fitness inheritance useful for real-world applications? Evolutionary Multi-Criterion Optimization. Second International Conference, EMO 2003, Springer. Lecture Notes in Computer Science. 2632:31–42
[68] Reynolds R G, Chung C-J (1997) A cultural algorithm framework to evolve multi-agent cooperation with evolutionary programming. Proceedings of the 6th International Conference on Evolutionary Programming VI, Springer-Verlag, 323–334

[69] Saleem S, Reynolds R (2000) Cultural algorithms in dynamic environments. Proceedings of the Congress on Evolutionary Computation, 1513–1520
[70] Reynolds R G, Chung C-J (1997) Knowledge-based self-adaptation in evolutionary programming using cultural algorithms. Proceedings of 1997 IEEE International Conference on Evolutionary Computation (ICEC 97), 71–76
[71] Reyes Sierra María M (2006) Use of coevolution and fitness inheritance for multiobjective particle swarm optimization. Computer Science Section, Department of Electrical Engineering, CINVESTAV-IPN, Mexico
[72] Giunta A A, Watson L (1998) A comparison of approximation modeling techniques: Polynomial versus interpolating models. AIAA, 47–58
[73] Jin R, Chen W, Simpsonand T W (2000) Comparative studies of metamodeling techniques under miltiple modeling criteria. AIAA, 2000–4801
[74] Goel T, Vaidyanathan R, Haftka R, Shyy W, Queipo N, Tucker K (2004) Response surface approximation of Pareto optimal front in multiobjective optimization. AIAA, 2004–4501

3

Evolutionary Multi-objective Rule Selection for Classification Rule Mining

Hisao Ishibuchi, Isao Kuwajima, Yusuke Nojima

Department of Computer Science and Intelligent Systems, Graduate School of Engineering, Osaka Prefecture University, 1-1 Gakuen-cho, Naka-ku, Sakai, Osaka 599-8531, Japan
hisaoi@cs.osakafu-u.ac.jp, kuwajima@ci.cs.osakafu-u.ac.jp, nojima@cs.osakafu-u.ac.jp

Summary. This chapter discusses the application of evolutionary multi-objective optimization (EMO) to classification rule mining. In the field of classification rule mining, classifiers are designed through the following two phases: rule discovery and rule selection. In the rule discovery phase, a large number of classification rules are extracted from training data. This phase is based on two rule evaluation criteria: support and confidence. An association rule mining technique such as Apriori is usually used to extract classification rules satisfying pre-specified threshold values of the minimum support and confidence. In some studies, EMO algorithms were used to search for Pareto-optimal rules with respect to support and confidence. On the other hand, a small number of rules are selected from the extracted rules to design an accurate and compact classifier in the rule selection phase. A heuristic rule sorting criterion is usually used for rule selection. In some studies, EMO algorithms were used for multi-objective rule selection to maximize the accuracy of rule sets and minimize their complexity. In this chapter, first we explain the above-mentioned two phases in classification rule mining. Next we explain the search for Pareto-optimal rules and the search for Pareto-optimal rule sets. Then we explain evolutionary multi-objective rule selection as a post processing procedure in the second phase of classification rule mining. A number of Pareto-optimal rule sets are found from a large number of candidate rules, which are extracted from training data in the first phase. Finally we show experimental results on some data sets from the UCI machine learning repository. Through computational experiments, we demonstrate that evolutionary rule selection can drastically decrease the number of extracted rules without severely degrading their classification accuracy. We also examine the relation between Pareto-optimal rules and Pareto-optimal rule sets.

3.1 Introduction

Data mining is a very active and rapidly growing research area in the field of computer science. The task of data mining is to extract useful knowledge for human users

H. Ishibuchi et al.: *Evolutionary Multi-objective Rule Selection for Classification Rule Mining*, Studies in Computational Intelligence (SCI) **98**, 47–70 (2008)
www.springerlink.com

from a large data set. Whereas the application of evolutionary computation to data mining is not always easy due to its heavy computation load especially in the case of a large data set (10; 11; 15), many evolutionary approaches have been proposed in the literature (3; 12; 35; 37; 39). Evolutionary multi-objective optimization (EMO) has also been applied to data mining in some studies (16; 17; 18; 21; 25; 26; 38). In the field of fuzzy logic, multi-objective formulations have frequently been used for knowledge extraction (6; 7; 23; 24; 29; 30; 41; 42) since the late 1990s (22). This is because the interpretability-accuracy tradeoff analysis is a very important research issue in the design of fuzzy rule-based systems (6; 7). Multi-objective formulations have also been used in non-fuzzy genetics-based machine learning (8; 31; 34).

Association rule mining (5) is one of the most well-known data mining techniques. form (5), association rules satisfying pre-specified threshold values of the minimum support and confidence are extracted from a data set. The application of association rule mining to classification problems is often referred to as classification rule mining or associative classification (32; 33; 36; 40). Classification rule mining usually consists of two phases: rule discovery and rule selection. In the rule discovery phase, a large number of classification rules are extracted from a data set using an association rule mining technique. Classification rules satisfying pre-specified threshold values of the minimum support and confidence are usually extracted from a data set. A part of extracted rules are selected to design an accurate and compact classifier in the rule selection phase using a heuristic rule sorting criterion.

Whereas the basic form of association rule mining is to extract association rules satisfying pre-specified threshold values of the minimum support and confidence (5), other rule evaluation measures have been proposed to qualify the *interestingness* or *goodness* of an association rule. Among them are gain, variance, chi-squared value, entropy gain, gini, laplace, lift, and conviction (9). It is shown in (9) that the best rule according to any of the above-mentioned measures is a Pareto-optimal rule with respect to support and confidence. Motivated by this study, the use of an EMO algorithm was proposed to search for Pareto-optimal rules with respect to support and confidence for partial classification (16; 17; 18; 38). Partial classification is the classification of a particular single class from all the other classes. Similar formulations to (16; 17; 18; 38) were used to search for Pareto-optimal association rules (21) and Pareto-optimal fuzzy association rules (30). EMO algorithms were also used to search for Pareto-optimal rule sets in classification rule mining (25; 26) where the accuracy of rule sets was maximized and their complexity was minimized. The same idea was used in the multi-objective design of fuzzy rule-based classifiers (22; 23; 29).

In this chapter, we examine the effect of evolutionary multi-objective rule selection on the accuracy and the complexity of selected rules through computational experiments on some well-known benchmark data sets from the UCI machine learning repository. We also examine the relation between Pareto-optimal rules and Pareto-optimal rule sets in classifier design. This examination is performed by depicting selected rules together with candidate rules in the support-confidence plane. Our interest is to examine whether selected rules in Pareto-optimal rule sets are close to the Pareto front with respect to support and confidence.

This chapter is organized as follows. First we explain some basic concepts in classification rule mining in Section 3.2. Next we explain two approaches in evolutionary multi-objective classification rule mining in Section 3.3. One approach handles each classification rule as an individual to search for Pareto-optimal rules. The other approach handles each rule set as an individual to search for Pareto-optimal rule sets. Then we explain evolutionary multi-objective rule selection as a post processing procedure in the rule selection phase of classification rule mining in Section 3.4. Pareto-optimal rule sets are found from a large number of candidate rules, which are extracted from a data set using an association rule mining technique in the rule discovery phase. Finally we report experimental results on some well-known benchmark data sets in Section 3.5. Experimental results demonstrate the effect of evolutionary multi-objective rule selection on the accuracy and the complexity of selected rules. The relation between Pareto-optimal rules and Pareto-optimal rule sets is also examined in Section 5. Section 6 concludes this chapter.

3.2 Classification Rule Mining

Let us assume that we have m training patterns $\mathbf{x}_p = (x_{p1}, x_{p2}, ..., x_{pn}), p = 1, 2, ..., m$ from M classes in an n-dimensional continuous pattern space where x_{pi} is the attribute value of the p-th training pattern for the i-th attribute. We denote the set of these m training patterns by D. For our pattern classification problem, we use classification rules of the following type:

$$\text{Rule } R_q : \text{If } x_1 \text{ is } A_{q1} \text{ and ... and } x_n \text{ is } A_{qn} \text{ then Class } C_q \text{ with } CF_q, \tag{3.1}$$

where R_q is the label of the q-th rule, $\mathbf{x} = (x_1, x_2, ..., x_n)$ is an n-dimensional pattern vector, A_{qi} is an antecedent interval for the i-th attribute, C_q is a class label, and CF_q is a rule weight (i.e., certainty grade). We denote the classification rule R_q in (3.1) as "$\mathbf{A}_q \Rightarrow C_q$" where $\mathbf{A}_q = (A_{q1}, A_{q2}, ..., A_{qn})$. Each antecedent condition "x_i is A_{qi}" in (3.1) means the inclusion relation "$x_i \in A_{qi}$". It should be noted that classification rules of the form in (3.1) do not always have n antecedent conditions. Some rules may have only a few conditions while others may have many conditions. That is, A_{qi} in (3.1) can be a *don't care* condition.

In the field of association rule mining, two rule evaluation measures called *support* and *confidence* have often been used (5; 9). Let us denote the support count of the classification rule $\mathbf{A}_q \Rightarrow C_q$ by $SUP(\mathbf{A}_q \Rightarrow C_q)$, which is the number of patterns compatible with both the antecedent part $\mathbf{A}_q$ and the consequent class C_q. $SUP(\mathbf{A}_q)$ and $SUP(C_q)$ are also defined in the same manner, which are the number of patterns compatible with $\mathbf{A}_q$ and C_q, respectively. The support of the classification rule $\mathbf{A}_q \Rightarrow C_q$ is defined as

$$Support(\mathbf{A}_q \Rightarrow C_q) = \frac{SUP(\mathbf{A}_q \Rightarrow C_q)}{|D|}, \tag{3.2}$$

where $|D|$ is the cardinality of the data set D (i.e., $|D| = m$). On the other hand, the confidence of $\mathbf{A}_q \Rightarrow C_q$ is defined as

$$Confidence(\mathbf{A}_q \Rightarrow C_q) = \frac{SUP(\mathbf{A}_q \Rightarrow C_q)}{SUP(\mathbf{A}_q)}. \tag{3.3}$$

The confidence is directly used as the rule weight in this chapter.

In partial classification (16; 17; 18; 38), the following measure called *coverage* is often used instead of the support:

$$Coverage(\mathbf{A}_q \Rightarrow C_q) = \frac{SUP(\mathbf{A}_q \Rightarrow C_q)}{SUP(C_q)}. \tag{3.4}$$

Since the consequent class is fixed in partial classification (i.e., since the denominator of (3.4) is constant), the maximization of the coverage is the same as that of the support.

In classification rule mining (32; 33; 36; 40), an association rule mining technique such as Apriori (5) is used in the rule discovery phase to efficiently extract classification rules satisfying pre-specified threshold values of the minimum support and confidence. These two parameters are assumed to be pre-specified by users. A part of extracted rules are selected to design an accurate and compact classifier in the rule selection phase.

Let S be a set of selected classification rules. That is, S is a classifier. When a new pattern $\mathbf{x}_p$ is to be classified by S, we choose a single winner rule with the maximum rule weight among compatible rules with $\mathbf{x}_p$ in S. The consequent class of the winner rule is assigned to $\mathbf{x}_p$. When multiple compatible rules with different consequent classes have the same maximum rule weight, the classification of $\mathbf{x}_p$ is rejected in evolutionary multi-objective rule selection in this chapter. Only when the accuracy of the finally obtained rule set is to be evaluated, we use random tie break among those classes with the same maximum rule weight in computational experiments.

3.3 Evolutionary Multi-objective Rule Mining

Evolutionary multi-objective techniques in classification rule mining can be roughly categorized into two approaches. In one approach, each rule is evaluated according to multiple rule evaluation criteria such as support and confidence. An EMO algorithm is used to search for Pareto-optimal rules. In the other approach, each rule set is evaluated according to multiple rule set evaluation criteria such as accuracy and complexity. An EMO algorithm is used to search for Pareto-optimal rule sets. In this section, we briefly explain these two approaches.

3.3.1 Techniques to Search for Pareto-Optimal Rules

It is shown in (9) that the set of Pareto-optimal rules with respect to support and confidence includes the best rule according to any of the following rule evaluation criteria: gain, variance, chi-squared value, entropy gain, gini, laplace, lift, and conviction. Thus it is an important research issue to search for Pareto-optimal rules with respect to support and confidence in association rule mining. The use of NSGA-II (2; 19) for this task was proposed by de la Iglesia et al. (16; 18) where they applied NSGA-II to the following two-objective optimization problem for partial classification.

$$\text{Maximize } \{Converge(R),\, Confidence(R)\}, \tag{3.5}$$

where R denotes a classification rule. It should be noted that the maximization of the coverage means that of the support since the consequent class is fixed in partial classification. The use of a dissimilarity measure between classification rules instead of a crowding measure in NSGA-II was examined in (17) in order to search for a set of Pareto-optimal rules with a large diversity. The Pareto-dominance relation in NSGA-II was modified in (38) in order to search for not only Pareto-optimal rules but also near Pareto-optimal rules.

Ghosh and Nath (21) used an EMO algorithm to search for Pareto-optimal association rules with respect to *confidence*, *comprehensibility* and *interestingness*. That is, association rule mining was formulated as a three-objective optimization problem in (21). A similar three-objective optimization problem was formulated in Kaya (30) where an EMO algorithm was used to search for Pareto-optimal fuzzy association rules with respect to *support*, *confidence* and *comprehensibility*.

3.3.2 Techniques to Search for Pareto-Optimal Rule Sets

In the rule selection phase, a part of extracted rules are selected using a heuristic rule sorting criterion in the rule selection phase to design an accurate and compact classifier. Evolutionary multi-objective rule selection was proposed in (25; 26) to search for Pareto-optimal rule sets with respect to accuracy and complexity in the rule selection phase of classification rule mining.

Evolutionary rule selection was originally proposed for the design of accurate and compact fuzzy rule-based classifiers in (28) where a weighted sum fitness function was used to maximize the classification accuracy and minimize the number of fuzzy rules. An EMO algorithm was used to search for Pareto-optimal fuzzy rule-based classifiers with respect to these two objectives in (22). The total number of antecedent conditions was introduced as the third objective in (23) to minimize not only the number of fuzzy rules but also their length while maximizing the classification accuracy of fuzzy rule-based classifies. The use of a memetic EMO algorithm was examined to search for Pareto-optimal fuzzy rule-based classifiers with respect to these three objectives in (29). Evolutionary multi-objective fuzzy rule selection techniques in these studies were used for non-fuzzy classification rule mining in (25; 26). They were also generalized as multi-objective techniques for the design of fuzzy rule-based systems (e.g., (27; 41; 42)).

3.4 Evolutionary Multi-objective Rule Selection

Let us assume that we have already extracted N classification rules in the rule discovery phase of classification rule mining. These N rules are used as candidate rules in the rule selection phase. Let S be a subset of the N candidate rules (i.e., S is a classifier). We use a binary string of length N to represent S where "1" and "0" mean the inclusion in S and the exclusion from S of the corresponding candidate rule.

As in our former studies (25; 26), we use the following three objectives:

$f_1(S)$: The number of correctly classified training patterns by S,
$f_2(S)$: The number of selected rules in S,
$f_3(S)$: The total number of antecedent conditions over selected rules in S.

The first objective is maximized while the second and third objectives are minimized. The third objective can be viewed as the minimization of the total rule length since the number of antecedent conditions of each rule is often referred to as the rule length.

3.4.1 Rule Selection

A simple method to find a single rule set S from the N candidate rules is to search for the optimal rule set with respect to the following weighted sum fitness function:

$$fitness(S) = w_1 \cdot f_1(S) - w_2 \cdot f_2(S) - w_3 \cdot f_3(S), \tag{3.6}$$

where $\mathbf{w} = (w_1, w_2, w_3)$ is a non-negative weight vector. Any optimization technique can be used to maximize the weighted sum fitness function in (3.6). We use the following evolutionary rule selection method in this chapter.

Evolutionary Rule Selection Algorithm
Step 1: Randomly generate N_{pop} binary strings of length N as an initial population where N_{pop} is a user-definable parameter called the population size.
Step 2: Iterate the following operations N_{pop} times to generate an offspring population with N_{pop} strings.

2.1: Select a pair of parent strings from the current population by binary tournament selection based on the weighted sum fitness function.

2.2: Recombine the selected parent strings to generate an offspring using uniform crossover. This operation is applied with a pre-specified crossover probability. When the crossover operation is not applied to the selected pair of parent strings, one of them is randomly chosen and handled as an offspring in the following steps.

2.3: Apply biased mutation to the offspring. This operation changes 0 to 1 with a small probability and 1 to 0 with a large probability to decrease the number of 1's (i.e., to decrease the number of selected rules) in the offspring.

2.4: Remove unnecessary rules from the offspring in a heuristic manner as explained later in this subsection.

Step 3: Select the best N_{pop} strings with respect to the weighted sum fitness function from the current and offspring populations.
Step 4: If a pre-specified termination condition is not satisfied, return to Step 2 with the best N_{pop} strings selected in Step 3 as the population in the next generation. Otherwise, choose the single best one from the N_{pop} strings and terminate the execution of the algorithm.

This evolutionary rule selection algorithm can be viewed as a single-objective genetic algorithm (SOGA) with the $(\mu + \lambda)$-ES generation update mechanism to optimize the weighted sum fitness function of the three objectives in (3.6). The number of offspring strings in Step 2 is the same as the population size (i.e., $\mu = \lambda = N_{\text{pop}}$).

Since we use the single winner-based method without random tie break to evaluate the accuracy of the rule set S in evolutionary rule selection as explained in Section 3.2, only a single rule is responsible for the classification of each training pattern. As a result, some rules may be used for the classification of no training patterns. Whereas the existence of such an unnecessary rule in the rule set S has no effect on the first term of the weighted sum fitness function, it deteriorates the second and third terms. Thus we remove from the offspring all the unnecessary rules responsible for the classification of no training patterns in Step 2.4. It should be noted that Step 2.4 can be viewed as a confidence-based heuristic rule pruning procedure since the choice of a single winner rule is based on the confidence (i.e., rule weight) of each rule. When multiple compatible rules with the same consequent class have the same maximum rule weight for a training pattern, we randomly choose one of those rules as a winner rule. On the other hand, the classification of a training pattern is rejected in the calculation of the weighted sum fitness function when multiple compatible rules with different consequent classes have the same maximum rule weight. It should be noted that random tie break is used in the latter case when the accuracy of the final result (i.e., the finally selected rule set) is evaluated after evolutionary rule selection. Random tie break is also used when the accuracy of candidate rules before evolutionary rule selection is evaluated for comparison.

We use the total number of iterations of the algorithm (i.e., the total number of generations) as the termination condition in this chapter. The best rule set among examined ones during the execution of our evolutionary rule selection algorithm is returned to users as the final result.

3.4.2 Accuracy-Complexity Tradeoff in Rule Selection

Since there exists a tradeoff relation between the accuracy of classifiers and their complexity, it is impossible to design an ideal classifier with the maximum accuracy and the minimum complexity by evolutionary rule selection. The solid curve in Figure 3.1 shows a typical relation between errors and complexity in classifier design. As shown by the solid curve, classification errors on training patterns monotonically decrease with the increase in the complexity of classifiers in many cases. A single point on this curve corresponds to the optimal rule set with respect to the weighted sum fitness function. The location of the optimal rule set depends on the specification of the weight vector $\mathbf{w} = (w_1, w_2, w_3)$. By increasing the value of the first weight w_1 for the accuracy, the location of the optimal rule set moves right along the solid curve in Figure 3.1.

On the other hand, classification errors on test patterns usually have a different characteristic feature from the accuracy-complexity tradeoff curve for training patterns. The dotted curve in Figure 3.1 shows a typical accuracy-complexity tradeoff relation for test patterns. That is, classification errors on test patterns first decrease then increase with the increase in the complexity of classifiers. Finding an optimal complexity with the minimum errors on test patterns (i.e., S^* in Figure 3.1 with the highest generalization ability) has been one of the main research issues in the machine learning community (1).

The optimal rule set S^* with the highest generalization ability in Figure 3.1 is not necessarily optimal in the data mining perspective especially when the complexity of S^* is high. Human users may prefer more interpretable rule sets with lower complexity to the optimal rule set S^* even if their classification accuracy is degraded (i.e., preferred rule sets by human users may lie to the left of S^* in Figure 3.1). Evolutionary multi-objective rule selection tries to find not only a single rule set but also a number of Pareto-optimal rule sets in order to meet various preferences of human users. That is, many points with different tradeoffs on the solid curve in Figure 3.1 are found by evolutionary multi-objective rule selection.

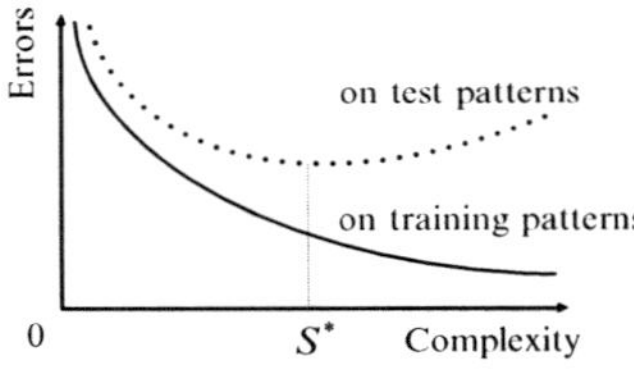

Fig. 3.1. Typical relations between classification errors and complexity of rule sets

3.4.3 Multi-objective Rule Selection

Since only a single rule set is obtained from a single run of the SOGA-based rule selection algorithm in Subsection 3.4.1, its multiple executions with different specifications of the weight vector are required to find a number of Pareto-optimal rule sets with different tradeoffs. This makes the search for Pareto-optimal rule sets a very time-consuming task. To perform multi-objective rule selection more efficiently, we can use evolutionary multi-objective optimization (EMO) algorithms instead of SOGA in Subsection 3.4.1. The use of EMO algorithms for multi-objective rule selection was first proposed for the multi-objective design of fuzzy rule-based classifiers with high accuracy and high interpretability in the late 1990s by Ishibuchi et al. (22).

As an EMO algorithm, we use NSGA-II (2; 19) to search for Pareto-optimal rule sets of the following three-objective rule selection problem:

$$\text{Maximize } f_1(S), \text{ minimize } f_2(S) \text{ and } f_3(S). \tag{3.7}$$

NSGA-II for multi-objective rule selection is the same as SOGA for rule selection in Subsection 3.4.1 except for the fitness evaluation mechanism for parent selection and generation update. Whereas the weighted sum fitness function was used in SOGA, such a scalarizing fitness function is not used in NSGA-II. Rule sets are evaluated using Pareto ranking and the concept of crowding (see (2; 19) for details of NSGA-II). The best rank is assigned to all rule sets that are not dominated by any other rule sets in the current population. The second rank is assigned to all rule sets that are not dominated by any other rule sets except for the best rank rule sets. In this manner, a different rank is assigned to each rule set using Pareto dominance relation. The rank of each rule set is used as the primary criterion for parent selection (i.e., Step 2.1 in Subsection 3.4.1) and generation update (i.e., Step 3). Among rule sets with the same rank, rule sets in less crowded regions of the objective space are viewed as being better than rule sets in more crowded regions. That is, a crowding measure is used as the secondary criterion for parent selection and generation update. See Deb (2) for various approaches to multi-objective fitness evaluation.

The final result of evolutionary multi-objective rule selection is not a single rule set but a number of non-dominated rule sets with respect to the three objectives in (3.7). This is the main characteristic feature of evolutionary multi-objective rule selection.

3.5 Computational Experiments

In this section, we demonstrate how SOGA and NSGA-II can decrease the number of extracted rules and their rule length without severely degrading their classification accuracy through computational experiments on some well-known benchmark data sets with continuous attributes in the UCI machine learning repository. We also examine the relation between Pareto-optimal rules and Pareto-optimal rule sets.

3.5.1 Conditions of Computational Experiments

We used five data sets in Table 3.1. We did not use incomplete patterns with missing values. All attribute values were handled as real numbers. The domain of each attribute was divided into multiple intervals using an optimal splitting method (20) based on the class entropy measure (4). Since the choice of an appropriate number of intervals for discretization is not easy, we simultaneously used four different partitions with two, three, four, and five intervals (i.e., 14 antecedent intervals in total for each attribute). As a result, various classification rules were examined and extracted in the rule discovery phase using overlapping antecedent intervals of various widths for each attribute.

Table 3.1. Data sets used in computational experiments

Data set	Attributes	Patterns	Classes
Breast W	9	683*	2
Glass	9	214	6
Heart C	13	297*	5
Iris	4	150	3
Wine	13	178	3

$*$ Incomplete patterns with missing values are not included.

We extracted classification rules with three or less antecedent conditions using pre-specified threshold values of the minimum support and confidence. This restriction on the number of antecedent conditions is to find rule sets with high interpretability (i.e., because it is difficult for human users to intuitively understand long classification rules with many antecedent conditions). We examined 4×4 combinations of the following four specifications of each threshold for the five data sets in Table 1:

Minimum confidence: 60%, 70%, 80%, 90%,
Minimum support: 4%, 6%, 8%, 10% (for the wine data set)
1%, 2%, 5%, 10% (for the other data sets).

All the extracted rules for each combination of the two threshold values were used as candidate rules in the rule selection phase.

3.5.2 Effect of Evolutionary Rule Selection

First we show some experimental results by SOGA to clearly demonstrate the effect of evolutionary rule selection on the number of selected rules and their classification accuracy. Our SOGA-based rule selection algorithm was executed with the following parameter values:

Weight vector: (2, 1, 1) and (10, 1, 1),
Population size: 200 strings,
Crossover probability: 0.9 (uniform crossover),
Mutation probability: 0.05 ($1 \rightarrow 0$),
$1/N$ ($0 \rightarrow 1$) where N is string length,
Termination condition: 1000 generations.

To evaluate the classification accuracy of classifiers (i.e., rule sets) before and after genetic rule selection, we iterated the two-fold cross-validation procedure with 50% training patterns and 50% test patterns five times for each data set. We report average results over its five iterations in this subsection. For comparison, we applied our genetic rule selection algorithm and the confidence-based heuristic rule pruning technique used in Step 2.4 to the same candidate rules in our computational experiments. In the following, we report experimental results for each data set.

Wisconsin breast cancer data: Figure 3.2 (a) shows the number of candidate rules extracted in the rule discovery step for the Wisconsin breast cancer data set. In Figure 3.2 (a), a large number of rules were extracted in the first step of classification rule mining. The number of extracted rules strongly depends on the two threshold values. The number of extracted rules in Figure 3.2 (a) was decreased by heuristic rule pruning in Figure 3.2 (b). The reduction in the number of rules was more drastic in Figure 3.2 (c) and Figure 3.2 (d) by genetic rule selection than Figure 3.2 (b) by heuristic rule pruning. Different weight vectors were used in Figure 3.2 (c) and Figure 3.2 (d). It should be noted that the vertical axis of each plot in Figure 3.2 has a different scale. For example, six or seven rules were selected in Figure 3.2 (c) from about 15000 candidate rules in Figure 3.2 (a) when the minimum support was specified as 0.01 (i.e., 1%).

Figure 3.2 demonstrates that our evolutionary rule selection algorithm is applicable even when the number of candidate rules is tens of thousands. This means that SOGA can handle binary strings whose length is tens of thousands. Many applications of genetic algorithms do not handle such a long string. Our experimental results in Figure 3.2 show high applicability of genetic algorithms to large-scale combinatorial optimization problems.

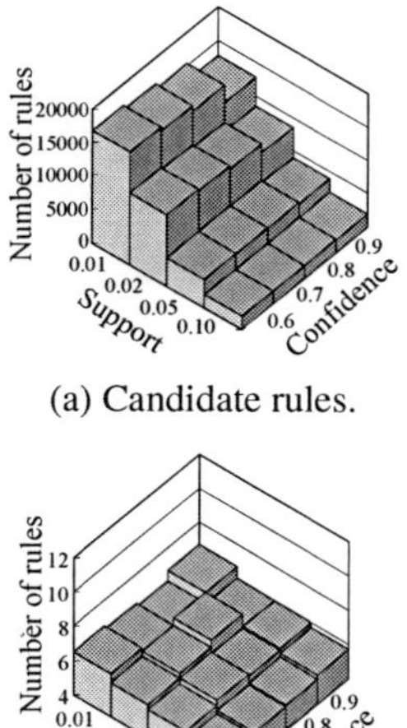

(a) Candidate rules.

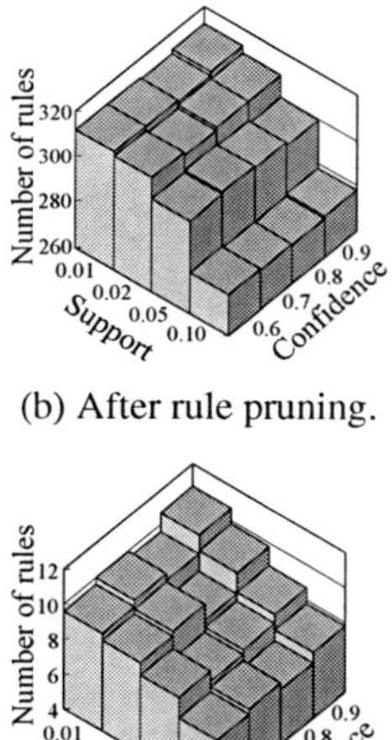

(b) After rule pruning.

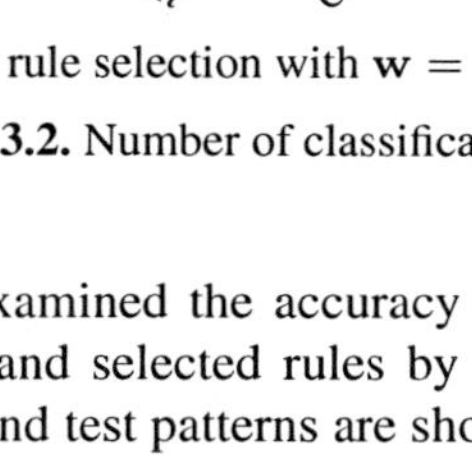

(c) After rule selection with $\mathbf{w} = (2, 1, 1)$.

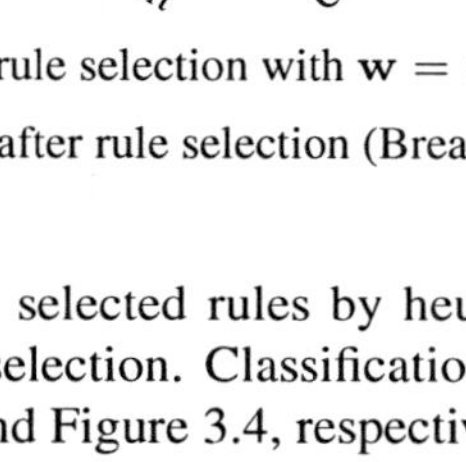

(d) After rule selection with $\mathbf{w} = (10, 1, 1)$.

Fig. 3.2. Number of classification rules before and after rule selection (Breast W)

We examined the accuracy of candidate rules, selected rules by heuristic rule pruning, and selected rules by evolutionary rule selection. Classification rates on training and test patterns are shown in Figure 3.3 and Figure 3.4, respectively. From Figure 3.3 (a) and Figure 3.3 (b), we can see that heuristic rule pruning does not change the accuracy of candidate rules on training patterns. This is because the winner rule for each training pattern is not removed by heuristic rule pruning. On the contrary, genetic rule selection improved the accuracy on training patterns when the weight vector (10, 1, 1) was used (compare Figure 3.3 (d) with Figure 3.3 (a)). This is because the combination of rules was globally optimized by genetic rule selection using a large weight value for the accuracy criterion (i.e., $w_1 = 10$). When the weight for the accuracy criterion was not so large (i.e., when $w_1 = 2$), the accuracy on training patterns was not improved in Figure 3.3 (c) from Figure 3.3 (a) except for the right-most four bars with the minimum support of 0.10. When $w_1 = 2$, the number of rules was drastically decreased in Figure 3.2 (c). We discuss the dependency of the final results of genetic rule selection on the specification of the weight vector in the next subsection.

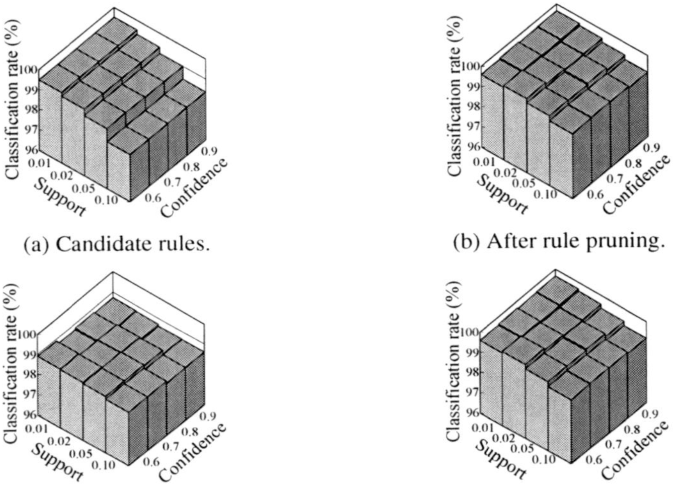

(a) Candidate rules. (b) After rule pruning.

(c) After rule selection with $\mathbf{w} = (2, 1, 1)$. (d) After rule selection with $\mathbf{w} = (10, 1, 1)$.

Fig. 3.3. Classification rates on training patterns (Breast W)

(a) Candidate rules. (b) After rule selection with $\mathbf{w} = (2, 1, 1)$.

Fig. 3.4. Classification rates on test patterns (Breast W)

The accuracy on test patterns is compared between candidate rules and selected rules in Figure 3.4. Due to the page limitation, we only show the accuracy of selected rules using the weight vector (2, 1, 1). We observe the deterioration by $1\% - 2\%$ in the generalization ability in Figure 3.4 (b). Such a small deterioration in the generalization ability can be viewed as the cost of a large decrease in the complexity of rule sets in Figure 3.2 (c). Similar results were obtained by the weight vector (10, 1, 1). We also observed a slight deterioration (less than 1% in many cases) in the generalization ability by heuristic rule pruning for the Wisconsin breast cancer data set.

In Figure 3.5, we show the average number of antecedent conditions in each rule. Evolutionary rule selection tends to choose more general rules with less antecedent conditions in Figure 3.5 (b) than heuristic rule pruning in Figure 3.5 (a). That is, more easily interpretable rules were found by evolutionary rule selection. For example, we can see from Figure 3.2 (c) and Figure 3.5 (b) that about six rules with one or two antecedent conditions were selected by evolutionary rule selection with the weight vector (2, 1, 1). Whereas we did not show experimental results, the average rule length of candidate rules before rule selection was about 3.

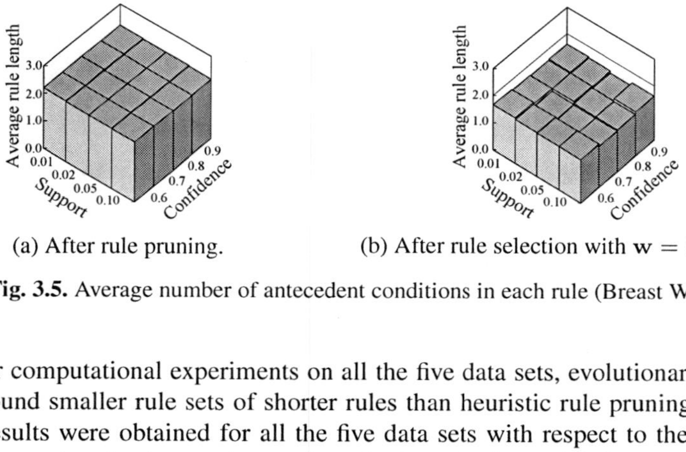

(a) After rule pruning. (b) After rule selection with $\mathbf{w} = (2, 1, 1)$.

Fig. 3.5. Average number of antecedent conditions in each rule (Breast W)

In our computational experiments on all the five data sets, evolutionary rule selection found smaller rule sets of shorter rules than heuristic rule pruning. That is, similar results were obtained for all the five data sets with respect to the effect of evolutionary rule selection on the number of selected rules and their average length. The effect of evolutionary rule selection on the training data accuracy was also very similar among the five data sets. That is, evolutionary rule selection improved the training data accuracy when the weight value for the accuracy criterion was large (i.e., when $w_1 = 10$). In the following, we only show the classification accuracy on test patterns for the other data sets because it was problem-dependent.

Glass data: In Figure 3.6, we show classification rates on test patterns. When the threshold value of the minimum confidence was 0.6, the generalization ability was improved by evolutionary rule selection. On the other hand, it was degraded when the threshold value of the minimum confidence was 0.9.

Cleveland heart disease data: Classification rates on test patterns are shown in Figure 3.7. The generalization ability was degraded by evolutionary rule selection in

(a) Candidate rules. (b) After rule selection with $\mathbf{w} = (2, 1, 1)$.

Fig. 3.6. Classification rates on test patterns (Glass)

all cases in Figure 3.7 (b). The deterioration is less severe when the threshold value of the minimum confidence was small.

(a) Candidate rules. (b) After rule selection with $\mathbf{w} = (2, 1, 1)$.

Fig. 3.7. Classification rates on test patterns (Heart C)

Iris data: Classification rates on test patterns are shown in Figure 3.8. An interesting observation in Figure 3.8 is that evolutionary rule selection improved the generalization ability in some cases with the minimum support 0.10. The deterioration in the generalization ability was less than 1% in the other cases.

(a) Candidate rules. (b) After rule selection with $\mathbf{w} = (2, 1, 1)$.

Fig. 3.8. Classification rates on test patterns (Iris)

Wine data: Classification rates on test patterns are shown in Figure 3.9. The generalization ability was degraded by evolutionary rule selection in all cases in

Figure 3.9. The deterioration in the generalization ability was less severe when the threshold value of the minimum support was large.

(a) Candidate rules.

(b) After rule selection with $\mathbf{w} = (2, 1, 1)$.

Fig. 3.9. Classification rates on test patterns (Wine)

3.5.3 Effect of Evolutionary Multi-objective Rule Selection

As we have already shown in the previous subsection (e.g., Figure 3.2 and Figure 3.3), different rule sets are obtained from the same candidate rules by evolutionary rule selection with different specifications of the weight vector. Using the Wisconsin breast cancer data set, we examined the dependency of the accuracy and the complexity of obtained rule sets on the specification of the weight vector. As in the previous subsection, we calculated the average classification rates on training and test patterns by iterating the two-fold cross-validation procedure five times. Candidate rules were extracted from training patterns by specifying the threshold values of the minimum support and confidence as 1% and 60%, respectively. The average number of extracted rules was 16828.

Using the same parameter specifications as the previous subsection except for the weight vector, we applied our evolutionary rule selection algorithm to the extracted candidate rules. Experimental results are summarized in Table 3.2 where eight specifications of the weight vector were examined.

We can observe the tradeoff relation between the complexity (i.e., the second and third columns) and the accuracy on training patterns (i.e., the fourth column) in Table 3.2. While the complexity of rule sets was decreased by the use of smaller values of w_1, their classification accuracy on training patterns was degraded. Moreover, a careful specification of the weight vector seems to be needed to maximize the generalization ability in Table 3.2.

Since only a single rule set is obtained from a single run of our SOGA-based rule selection algorithm, its multiple executions with different specifications of the weight vector are required to find a number of rule sets with different tradeoffs as shown in Table 3.2. This makes tradeoff analysis a very time-consuming task. To perform tradeoff analysis more efficiently, we can use evolutionary multi-objective rule selection. As we have already explained, we use NSGA-II for evolutionary multi-objective rule selection.

Table 3.2. Results with different specifications of the weight vector (Breast W)

Weights (w_1, w_2, w_3)	Number of rules	Length of rules	Training accuracy	Test accuracy
(100, 1, 1)	14.6	2.0	99.8	94.1
(50, 1, 1)	10.8	1.9	99.8	94.0
(0, 1, 1)	10.0	1.9	99.7	94.8
(10, 1, 1)	9.6	2.0	99.6	95.2
(5, 1, 1)	9.2	1.9	99.6	94.3
(2, 1, 1)	6.8	1.6	98.8	95.4
(1, 1, 1)	4.8	1.4	97.9	94.6
(0.5, 1, 1)	0.0	0.0	0.0	0.0

We applied evolutionary multi-objective rule selection to the Wisconsin breast cancer data set in the following manner. First the given 683 patterns were randomly divided into 342 training patterns and 341 test patterns. Next candidate rules were extracted from the 342 training patterns using the minimum support 0.01 and the minimum confidence 0.6. As a result, 17070 classification rules were extracted. Then NSGA-II was applied to the extracted classification rules. From its single run, 20 non-dominated rule sets were obtained. Finally each of the obtained rule sets was evaluated for the training and test patterns. The classification rates of obtained rule sets are shown in Figure 3.10 (a) for the training patterns and Figure 3.10 (b) for the test patterns. Some of the obtained rule sets (i.e., a rule set with only a single rule) are not shown because their classification rates are out of the range of the vertical axis of each plot in Figure 3.10. We can observe a clear tradeoff relation between the number of selected rules and the classification rates on the training patterns in Figure 3.10 (a). A similar tradeoff relation is also observed for the test patterns in Figure 3.10 (b).

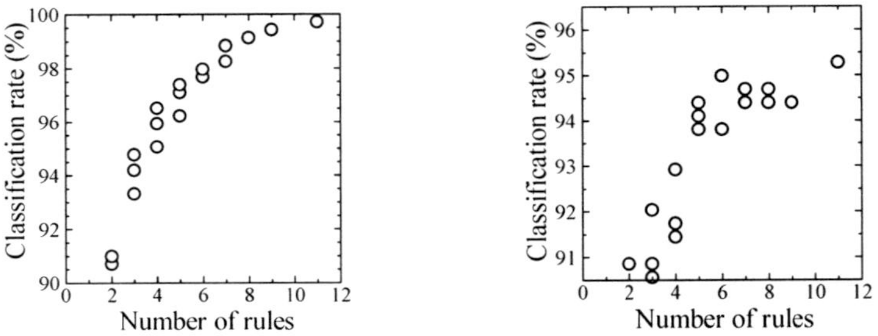

(a) Training data accuracy. (b) Test data accuracy.

Fig. 3.10. Results by evolutionary multi-objective rule selection (Breast W)

It should be noted that all rule sets in Figure 3.10 were obtained by a single run of evolutionary multi-objective rule selection (i.e., NSGA-II). If we try to find a similar number of rule sets using the SOGA-based rule selection algorithm, we need at least 20 runs of SOGA with different specifications of the weight vector. This is a clear advantage of EMO algorithms over SOGAs in evolutionary rule selection. Since different rule sets are not always obtained from different weight vectors, the specification of the weight vector is very difficult in the search for rule sets with large diversity. On the other hand, most EMO algorithms such as NSGA-II have diversity-maintenance mechanisms.

We performed similar computational experiments on the other four data sets. That is, evolutionary multi-objective rule selection was performed using 50% of each data set as training patterns. The generalization ability of selected rule sets was examined using the remaining 50% of each data set as test patterns. Experimental results on the other four data sets are shown in Figures 3.11-3.14. While we observed very similar tradeoff relations between the accuracy on training patterns and the number of selected rules for all the five data sets in Figures 3.10-3.14 (a), we obtained different results on test patterns as shown in Figures 3.10-3.14 (b). For example, we observe a clear deterioration in the generalization ability due to the increase in the number of selected rules in Figure 3.12 (b) for the Cleveland heart disease data set.

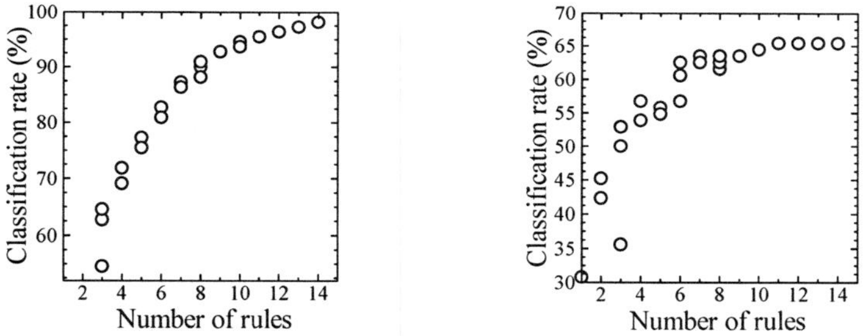

(a) Training data accuracy. (b) Test data accuracy.

Fig. 3.11. Results by evolutionary multi-objective rule selection (Glass)

3.5.4 Relation Between Pareto-optimal Rules and Pareto-optimal Rule Sets

Finally we examine the relation between Pareto-optimal rules and Pareto-optimal rule sets. More specifically, we examine whether Pareto-optimal rules were selected in each rule set in Figures 3.10-3.14 by depicting the selected rules in the support-confidence plane. In Figure 3.15, we show candidate rules and Pareto-optimal rules for the Glass data set and the Cleveland heart disease data set. In each plot, small circles (i.e., dots) are candidate rules. Pareto-optimal rules are shown by open circles. Since Pareto-optimal rules are defined by support maximization and confidence maximization, they locate along the upper-right limit of candidate rules in the support-confidence plane.

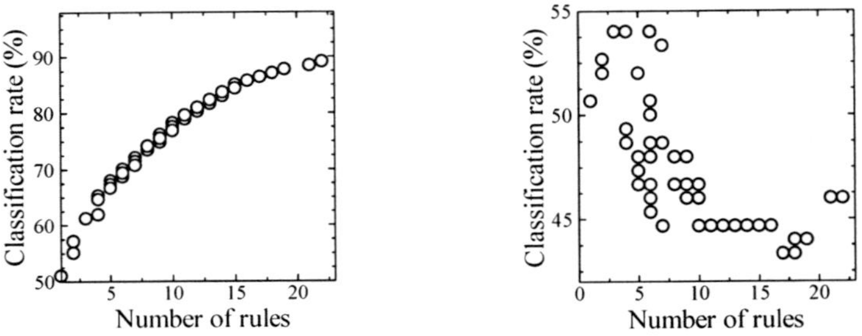

(a) Training data accuracy. (b) Test data accuracy.

Fig. 3.12. Results by evolutionary multi-objective rule selection (Heart C)

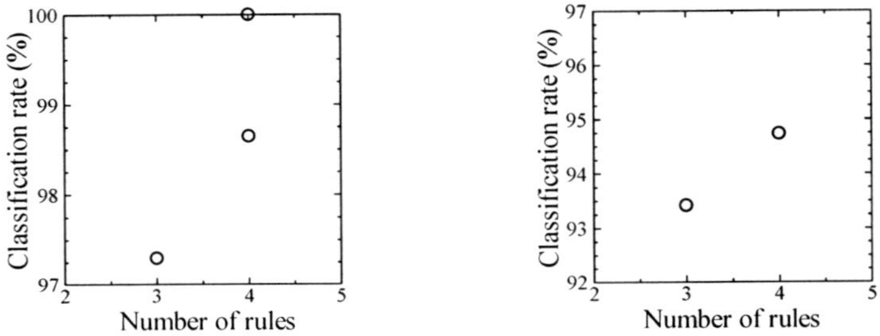

(a) Training data accuracy. (b) Test data accuracy.

Fig. 3.13. Results by evolutionary multi-objective rule selection (Iris)

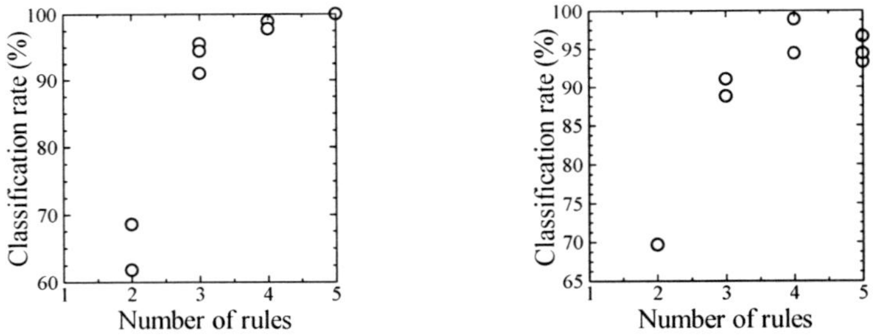

(a) Training data accuracy. (b) Test data accuracy.

Fig. 3.14. Results by evolutionary multi-objective rule selection (Wine)

In Figure 3.16 (b)-(d), we show the locations of the selected rules in the three rule sets in Figure 3.16 (a) for the glass data set. Experimental results on the Cleveland heart disease data set are shown in Figure 3.17 in the same manner. We can see from these figures that dominated rules were often selected in Pareto-optimal rule sets. We obtained the same observation from experimental results on the other three data sets.

Whereas selected rules are often far from Pareto-optimal rules in Figure 3.16 and Figure 3.17, they are usually very close to class-wise Pareto-optimal rules. For

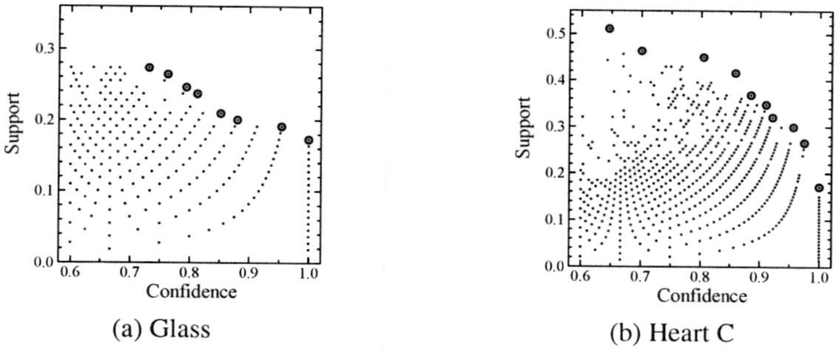

Fig. 3.15. Candidate rules (small circles) and Pareto-optimal rules (open circles)

example, the candidate rules and the selected rules in Figure 3.17 (c) with different consequent classes are separately depicted in Figure 3.18. As shown in this figure, selected rules are class-wise Pareto-optimal or very close to class-wise Pareto-optimal rules in many cases.

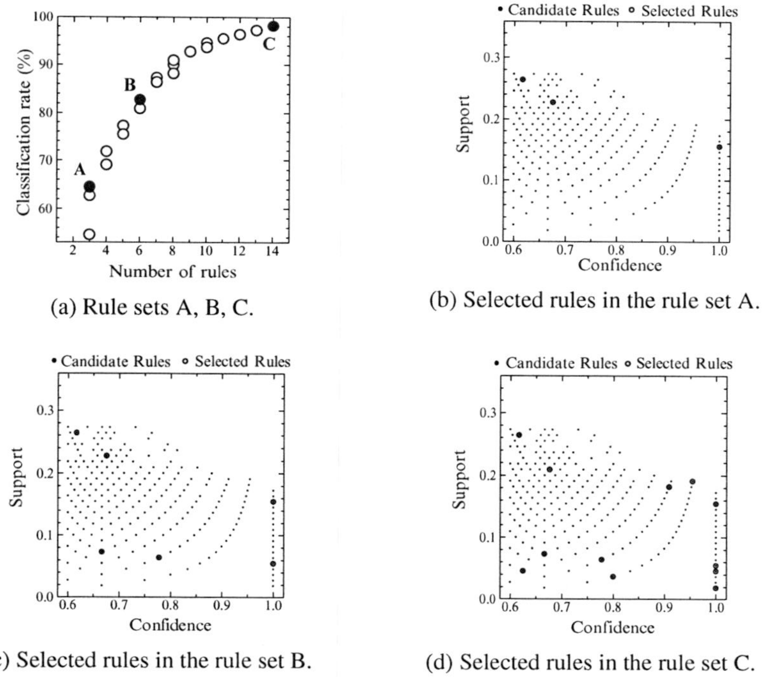

Fig. 3.16. Locations of selected rules in the support-confidence plane (Glass)

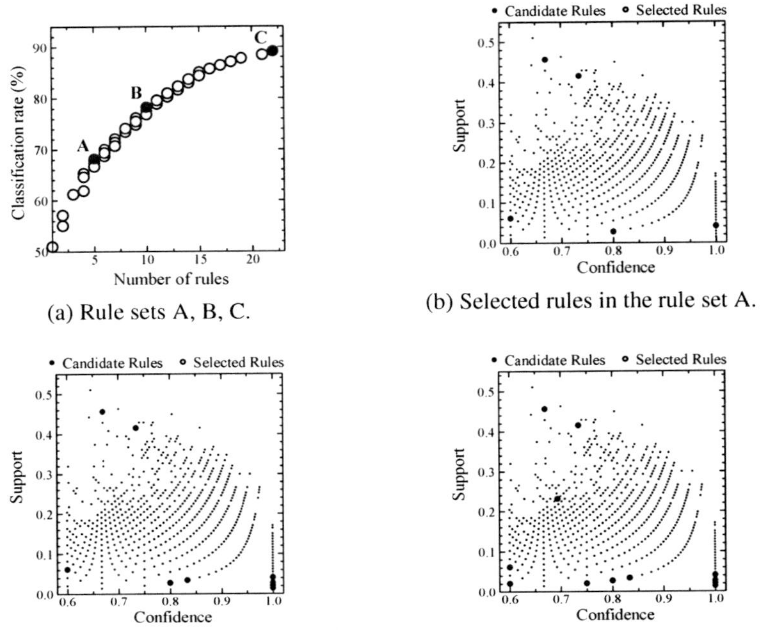

(a) Rule sets A, B, C.

(b) Selected rules in the rule set A.

(c) Selected rules in the rule set B.

(d) Selected rules in the rule set C.

Fig. 3.17. Locations of selected rules in the support-confidence plane (Heart C)

3.6 Conclusions

In this chapter, first we explained two approaches in evolutionary multi-objective classification rule mining. One is to search for Pareto-optimal rules and the other is to search for Pareto-optimal rule sets. Next we demonstrated that SOGA-based rule selection can significantly decrease the number of extracted rules without severely degrading their classification accuracy. It was also shown that SOGA-based rule selection can be applied to tens of thousands of candidate rules. Then we showed the accuracy-complexity tradeoff relation of selected rule sets by evolutionary multi-objective rule selection. One advantage of EMO-based rule selection over SOGA-based one is that a large number of Pareto-optimal (or near Pareto-optimal) rule sets can be obtained by a single run of an EMO algorithm while multiple runs of SOGA is needed. Finally we examined the relation between Pareto-optimal rules and Pareto-optimal rule sets. It was shown that dominated rules were often selected in Pareto-optimal rule sets. They were, however, very close to class-wise Pareto-optimal rules in many cases.

This work was partially supported by Grant-in-Aid for Scientific Research on Priority Areas (18049065) and for Scientific Research (B) (17300075).

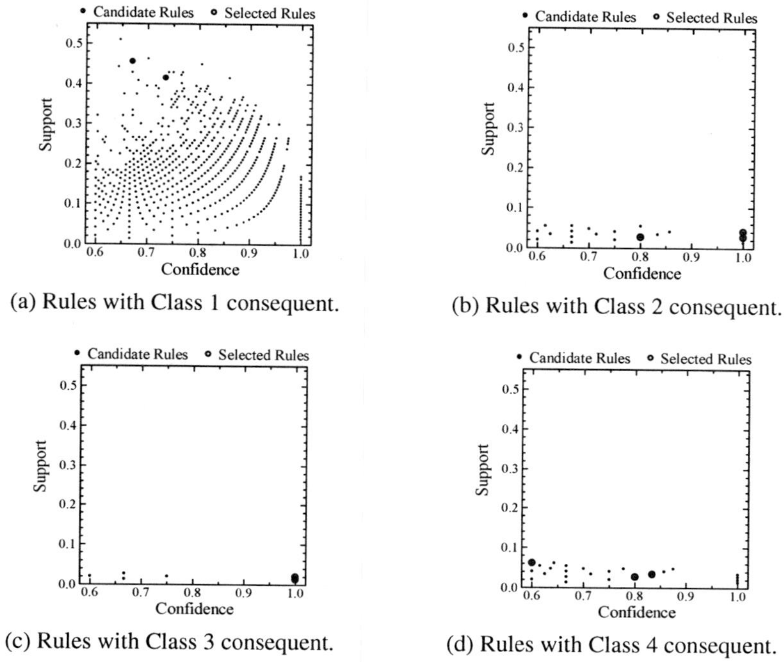

(a) Rules with Class 1 consequent. (b) Rules with Class 2 consequent.

(c) Rules with Class 3 consequent. (d) Rules with Class 4 consequent.

Fig. 3.18. Selected rules and candidate rules with the same consequent class

References

[1] Cherkassky V, Mulier F (1998) Learning from data: concept, theory, and methods. John Wiley & Sons, New York

[2] Deb K (2001) Multi-objective optimization using evolutionary algorithms. John Wiley & Sons, Chichester

[3] Freitas A A (2002) Data mining and knowledge discovery with evolutionary algorithms. Springer, Berlin

[4] Quinlan J R (1993) C4.5: Programs for machine learning. Morgan Kaufmann, San Mateo

[5] Agrawal R, Mannila H, Srikant R, Toivonen H, Verkamo A I (1996) Fast discovery of association rules. In Fayyad U M, Piatetsky-Shapiro G, Smyth P, Uthurusamy R (eds) Advances in Knowledge Discovery and Data Mining. AAAI Press, 307–328

[6] Casillas J, Cordon O, Herrera F, Magdalena L (2003) (eds) Interpretability issues in fuzzy modeling. Springer, Berlin

[7] Casillas J, Cordon O, Herrera F, Magdalena L (2003) (eds) Accuracy improvements in linguistic fuzzy modeling. Springer, Berlin

[8] Jin Y (2006) (eds) Multi-objective machine learning. Springer, Berlin

[9] Bayardo Jr R J, Agrawal R (1999) Mining the most interesting rules. Proc. of 5th ACM SIGKDD International Conference on Knowledge Discovery and Data Mining, 145–153

[10] Cano J R, Herrera F, Lozano M (2005) Stratification for scaling up evolutionary prototype selection. Pattern Recognition Letters 26:953–963

[11] Cano J R, Herrera F, Lozano M (2006) On the combination of evolutionary algorithms and stratified strategies for training set selection in data mining. Applied Soft Computing 6: 323–332

[12] Chiu C C, Hsu P L (2005) A Constraint-based genetic algorithm approach for mining classification rules. IEEE Transactions on Systems, Man, and Cybernetics: Part C - Applications and Reviews 35: 205–220

[13] Coenen F, Leng P (2005) Obtaining best parameter values for accurate classification. Proc. of 5th IEEE International Conference on Data Mining, 549–552

[14] Coenen F, Leng P, Zhang L (2005) Threshold tuning for improved classification association rule mining. Lecture Notes in Artificial Intelligence, Vol. 3518: Advances in Knowledge Discovery and Data Mining - PAKDD 2005. Springer, Berlin, 216–225

[15] Curry R, Heywood M I (2004) Towards efficient training on large datasets for genetic programming. Lecture Notes in Artificial Intelligence, Vol. 3060: Advances in Artificial Intelligence - Canadian AI 2004. Springer, Berlin, 161–174

[16] de la Iglesia B, Philpott M S, Bagnall A J, Rayward-Smith V J (2003) Data mining rules using multi-objective evolutionary algorithms. Proc. of 2003 Congress on Evolutionary Computation, 1552–1559

[17] de la Iglesia B, Reynolds A, Rayward-Smith V J (2005) Developments on a multi-objective metaheuristic (MOMH) Algorithm for finding interesting sets of classification rules. Lecture Notes in Computer Science, Vol. 3410: Evolutionary Multi-Criterion Optimization - EMO 2005. Springer, Berlin, 826–840

[18] de la Iglesia B, Richards G, Philpott M S, Rayward-Smith V J (2006) The application and effectiveness of a multi-objective metaheuristic algorithm for partial classification. European Journal of Operational Research 169: 898–917

[19] Deb K, Pratap A, Agarwal S, Meyarivan T (2002) A fast and elitist multiobjective genetic algorithm: NSGA-II. IEEE Transactions on Evolutionary Computation 6:182–197

[20] Elomaa T, Rousu J (1999) General and efficient multisplitting of numerical attributes. Machine Learning 36:201–244

[21] Ghosh A, Nath B T (2004) Multi-objective rule mining using genetic algorithms. Information Sciences 163:123–133

[22] Ishibuchi H, Murata T, Turksen I B (1997) Single-objective and Two-objective genetic algorithms for selecting linguistic rules for pattern classification problems. Fuzzy Sets and Systems 89:135–150

[23] Ishibuchi H, Nakashima T, Murata T (2001) Three-objective genetics-based machine learning for linguistic rule extraction. Information Sciences 136:109–133

[24] Ishibuchi H, Nakashima T, Nii M (2004) Classification and modeling with linguistic information granules: Advanced approaches to linguistic data mining. Springer, Berlin
[25] Ishibuchi H, Namba S (2004) Evolutionary multiobjective knowledge extraction for high-dimensional pattern classification problems. Lecture Notes in Computer Science, Vol. 3242: Parallel Problem Solving from Nature - PPSN VIII. Springer, Berlin, 1123–1132
[26] Ishibuchi H, Nojima Y (2005) Accuracy-complexity tradeoff analysis by multiobjective rule selection. Proc. of ICDM 2005 Workshop on Computational Intelligence in Data Mining, 39–48
[27] Ishibuchi H, Nojima Y (2006) Analysis of interpretability-accuracy tradeoff by multiobjective fuzzy genetics-based machine learning. International Journal of Approximate Reasoning (in press)
[28] Ishibuchi H, Nozaki K, Yamamoto N, Tanaka H (1995) Selecting fuzzy if-then rules for classification problems using genetic algorithms. IEEE Transactions on Fuzzy Systems 3:260–270
[29] Ishibuchi H, Yamamoto T (2004) Fuzzy rule selection by multi-objective genetic local search algorithms and rule evaluation measures in data mining. Fuzzy Sets and Systems 141:59–88
[30] Kaya M (2006) Multi-objective genetic algorithm based approaches for mining optimized fuzzy association rules. Soft Computing 10:578–586
[31] Kupinski M A, Anastasio M A (1999) Multi-objective genetic optimization of diagnostic classifiers with implications for generating receiver operating characteristic curve. IEEE Transactions on Medical Imaging 18:675–685
[32] Li W, Han J, Pei J (2001) CMAR: accurate and efficient classification based on multiple class-association rules. Proc. of 1st IEEE International Conference on Data Mining, 369–376
[33] Liu B, Hsu W, Ma Y (1998) Integrating classification and association rule mining. Proc. of 4th International Conference on Knowledge Discovery and Data Mining, 80–86
[34] Llora X, Goldberg D E (2003) Bounding the effect of noise in multi-objective learning classifier systems. Evolutionary Computation 11:278–297
[35] Mitra S, Pal S K, Mitra P (2002) Data mining in soft computing framework: A survey. IEEE Transactions on Neural Networks 13:3–14
[36] Mutter S, Hall M, Frank E (2004) Using classification to evaluate the output of confidence-based association rule mining. Lecture Notes in Artificial Intelligence, Vol. 3339: Advances in Artificial Intelligence - AI 2004. Springer, Berlin, 538–549
[37] Pal S K, Talwar V, Mitra P (2002) Web mining in soft computing framework: Relevance, state of the art and future directions. IEEE Transactions on Neural Networks 13:1163–1177
[38] Reynolds A, de la Iglesia B (2006) Rule induction using multi-objective metaheuristics: Encouraging rule diversity. Proc. of 2006 International Joint Conference on Neural Networks, 6375–6382

[39] Tan K C, Yu Q, Lee T H (2005) A distributed evolutionary classifier for knowledge discovery in data mining. IEEE Transactions on Systems, Man, and Cybernetics: Part C - Applications and Reviews 35:131–142

[40] Thabtah F, Cowling P, Hammoud S (2006) Improving rule sorting, predictive accuracy and training time in associative classification. Expert Systems with Applications 31:414–426

[41] Wang H, Kwong S, Jin Y, Wei W, Man K F (2005) Multi-objective hierarchical genetic algorithm for interpretable fuzzy rule-based knowledge extraction. Fuzzy Sets and Systems 149:149–186

[42] Wang H, Kwong S, Jin Y, Wei W, Man K F (2005) Agent-based evolutionary approach for interpretable rule-based knowledge extraction. IEEE Transactions on Systems, Man, and Cybernetics: Part C - Applications and Reviews 35:143–155

4

Rule Extraction from Compact Pareto-optimal Neural Networks

Yaochu Jin, Bernhard Sendhoff, Edgar Körner

Honda Research Institute Europe
63073 Offenbach/Main, Germany
yaochu.jin@honda-ri.de

Summary. Rule extraction from neural networks is a powerful tool for knowledge discovery from data. In order to facilitate rule extraction, trained neural networks are often pruned so that the extracted rules are understandable to human users. This chapter presents a method for extracting interpretable rules from neural networks that are generated using an evolutionary multi-objective algorithm. In the algorithm, the accuracy on the training data and the complexity of the neural networks are minimized simultaneously. Since there is a tradeoff between accuracy and complexity, a number of Pareto-optimal neural networks, instead of one single optimal neural network, are obtained. We show that the Pareto-optimal networks with a minimal degree of complexity are often interpretable in that understandable logic rules can be extracted from them straightforwardly. The proposed approach is verified on two benchmark problems.

4.1 Introduction

Knowledge acquired by neural networks is not easily understandable to human beings since the acquired knowledge is distributed among the weights and nodes of the neural networks. Many efforts have been made to extract symbolic or fuzzy rules from trained neural networks to gain a deep insight into the neural networks (1; 2; 4; 5; 6), which is of particular importance when the neural networks are employed for critical applications, or when neural networks are used for knowledge discovery from data (7).

As indicated in (1), two main issues must be taken into account in extracting rules from trained neural networks. First, the extracted rules should contain the same information that is learned by the original neural network. No significant information loss should occur, nor should additional information be introduced into the rules. Second, the extracted rules should be understandable to human users. As discussed in (3; 8; 9), one important aspect that is closely related to the interpretability of the extracted rules is the complexity of the extracted rules, including the number of rules and the number of premises in the rules. It is suggested that for the rules to

Y. Jin et al.: *Rule Extraction from Compact Pareto-optimal Neural Networks*, Studies in Computational Intelligence (SCI) **98**, 71–90 (2008)
www.springerlink.com

be easily understandable, the number of rules should be limited, and the number of premises should be small (usually less than 10). To improve the comprehensibility of the extracted rules, it is a common practice to prune the neural networks using regularization techniques.

Two basic approaches, namely, the de-compositional approach and the pedagogical approach, have been developed for extracting rules from trained neural networks (1). The de-compositional approach translates each node in the neural network into a rule, in which the inputs of the node are the premises of the rule, and the output is the consequence of the rule. The pedagogical approach, by contrast, treats the neural network as a blackbox and considers rule extraction as a learning process.

This chapter presents an approach to extracting interpretable rules from neural networks that are generated using an evolutionary multi-objective approach. The basic idea is to generate a number of Pareto-optimal neural networks that reflect a tradeoff between accuracy and complexity. Multiple neural networks of a variety of model complexities, instead of either a single signal-type or symbol-type model, will be generated using a multi-objective evolutionary algorithm combined with a local search, where accuracy and complexity serve as two conflicting objectives. It has been shown in (10; 11) that by analyzing the Pareto front, we are able to identify neural networks that are interpretable and those that are able to generalize on unseen data.

The use of evolutionary multi-objective algorithms (15; 16) for addressing machine learning problems has attracted increasing interest over the past few years. On the one hand, evolutionary algorithms are well suited for solving multi-objective problems mainly because they are population based search methods, which are able to achieve an approximation of the Pareto-front in one single run. On the other hand, machine learning problems are inherently multi-objective problems, where tradeoffs between accuracy and complexity, between accuracy and diversity, between accuracy and interpretability, and between stability and plasticity have to be taken into account. The marriage of evolutionary multi-objective optimization and machine learning has shown to be very fruitful, see (12) for a complete and updated overview of the research results.

Section 4.2 discusses very briefly the existing methods for controlling model complexity in the context of model selections in machine learning. Methods for extracting rules from neural networks are introduced. Section 4.3 shows that any formal neural network regularization method that simultaneously minimizes the accuracy and complexity can be treated as a multi-objective optimization problem. The details of the evolutionary multi-objective algorithm, together with the local search method will be provided in Section 4.4. Two illustrative examples are given in Section 4.5, where a number of Pareto-optimal neural networks are generated using the evolutionary multi-objective optimization algorithm. It will be shown that among the models generated by the multi-objective evolutionary algorithm, interpretable logic rules can be extracted from the compact Pareto-optimal neural networks.

4.2 Rule Extraction and Complexity Control

4.2.1 Model Selection in Machine Learning

By model selection, we mean to choose the best model for a set of given data, assuming that a number of models is available. Here, the meaning of "best" needs to be further explained. In general, a model is the best if the prediction error on the unseen data is minimal. Several criteria have been proposed based on the Kullback-Leibler Information Criterion (14). Two most popular criteria are Akaike's Information Criterion (AIC) and Bayesian Information Criterion (BIC). For example, model selection according to the AIC is to minimize the following criterion:

$$AIC = -2\ log(\mathcal{L}(\theta|y,\ g) + 2\ K, \tag{4.1}$$

where, $\mathcal{L}(\theta|y,\ g)$ is the maximized likelihood for data y given a model g with model parameter θ, K is the number of effective parameters of g.

The two terms in in Equation (4.1) indicate that model selection has to deal with two criteria. The first criterion minimizes the approximation error on the training data, whereas the second one minimizes the complexity of the model. There is a conflict between the two objectives, i.e., we cannot minimize the two objectives simultaneously. Usually, the smaller the error on the training data, the larger the complexity of the neural network. Obviously, a trade-off between accuracy and model complexity has to be handled in model selection.

Consequently, model selection criteria have often been employed to control the complexity of models to a desired degree. This approach is usually known as regularization in the neural network learning (13). The main purpose of neural network regularization is to avoid the overfitting of the training data by means of controlling the complexity of the neural network. In this way, the generalization capability of a neural network is improved. By generalization, it is meant that a trained neural network should perform well not only on the training data, but also on unseen data.

4.2.2 Complexity Reduction in Rule Extraction

Generally, neural networks are difficult to understand for human users. Due to this reason, many efforts have been made to extract symbolic or fuzzy rules from trained neural network (1; 2; 4). Two assumptions are often made during rule extraction from trained neural networks. First, units in the neural network are either maximally active or inactive. To meet this requirement, regularization techniques such as structural regularization (24), weight sharing (3) or network pruning (31) are usually implemented before rule extraction, which in effect reduces the complexity of neural networks and thus also contributes to the comprehensibility of the extracted rules. The complexity reduction procedure prior to rule extraction is also termed skeletonization (18). The second assumption is that a label, or in other words, a meaning needs to be associated with each unit. This is of less concern if rules are extracted from the output neurons.

Rule extraction from trained neural networks can be seen as a model selection process that trades off between accuracy and interpretability, where preference is put on the interpretability of the model. Thus, the trade-off between accuracy and complexity in model selection also reflects a trade-off between accuracy and interpretability.

Existing methods for extracting symbolic rules from trained neural network can largely be divided into three steps: neural network training, network skeletonization, and rule extraction (4).

4.3 Multi-objective Optimization Approach to Complexity Reduction

4.3.1 Neural Network Regularization

Neural network regularization can be realized by including an additional term that reflects the model complexity in the cost function of the training algorithm:

$$J = E + \lambda \Omega, \tag{4.2}$$

where E is the approximation error on the training data, Ω is the regularization term representing the complexity of the network model, and λ is a hyperparameter that controls the strength of the regularization. The most common error function in training or evolving neural networks is the mean squared error (MSE):

$$E = \frac{1}{N} \sum_{i=1}^{N} (y^d(i) - y(i))^2, \tag{4.3}$$

where N is the number of training samples, $y^d(i)$ is the desired output of the i-th sample, and $y(i)$ is the network output for the i-th sample. For the sake of clarity, we assume that the neural network has only one output. Refer to (13) for other error functions, such as the Minkowski error or cross-entropy.

Several measures have also been suggested for denoting the model complexity Ω. A most popular regularization term is the squared sum of all weights of the network:

$$\Omega = \frac{1}{2} \sum_{k} w_k^2, \tag{4.4}$$

where k is an index summing up all weights. This regularization method has been termed *weight decay*.

One weakness of the weight decay method is that it is not able to drive small irrelevant weights to zero, when gradient-based learning algorithms are employed, which may result in many small weights (28). An alternative is to replace the squared sum of the weights with the sum of absolute value of the weights:

$$\Omega = \sum_{i} |w_i|. \tag{4.5}$$

It has been shown that this regularization term is able to drive irrelevant weights to zero (26).

Note, however, that the weakness of weight decay is more related to the learning method (e.g., a gradient-based learning algorithm) other than the complexity measure itself. Different to the conclusions reported (26), where a gradient-based learning method has been used, it has been shown in (25) that regularization using the sum of squared weights is able to change (reduce) the structure of neural networks as efficiently as using the sum of absolute weights, when an evolutionary algorithm is employed to minimize the structure of neural networks.

A more direct measure for model complexity of neural networks is the number of connections contained in the neural network:

$$\Omega = \sum_i \sum_j c_{ij}, \tag{4.6}$$

where c_{ij} equals 1 denotes that there is connection from neuron j to neuron i, and not if $c_{ij} = 0$. It should be noticed that the above complexity measure is not generally applicable to gradient-based learning methods.

4.3.2 Multi-objective Optimization Approach to Regularization

It is quite straightforward to notice that neural network regularization in Equation (4.2) can be reformulated as a bi-objective optimization problem:

$$\min \{f_1, f_2\} \tag{4.7}$$

$$f_1 = E, \tag{4.8}$$

$$f_2 = \Omega, \tag{4.9}$$

where E is defined in Equation (4.3), and Ω is one of the regularization terms defined in Equation (4.4), (4.5), or (4.6).

It is noticed that regularization is traditionally formulated as a single objective optimization problem as in Equation (4.2) rather than a multi-objective optimization problem as in Equation (4.7). In our opinion, this tradition can be mainly attributed to the fact that traditional gradient-based learning algorithms are not able to solve multi-objective optimization problems.

4.4 Evolutionary Multi-objective Optimization of Neural Networks

4.4.1 Coding the Structure and Parameters of Neural Networks

A connection matrix and a weight matrix are employed to describe the structure and the weights of the neural networks, see Figure 4.1. The connection matrix specifies the structure of the network, whereas the weight matrix determines the strength

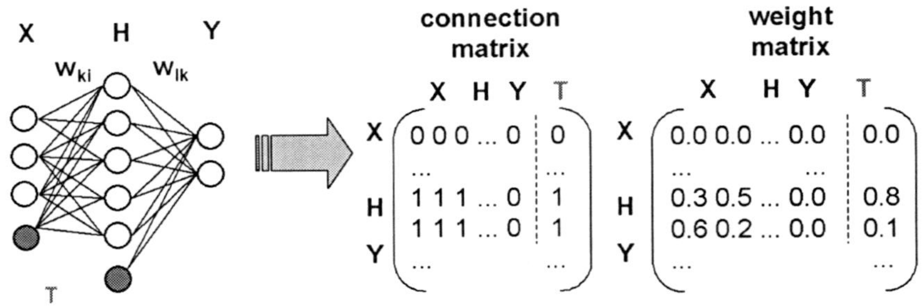

Fig. 4.1. Coding of the structure and parameters of neural networks using a connection matrix and a weight matrix

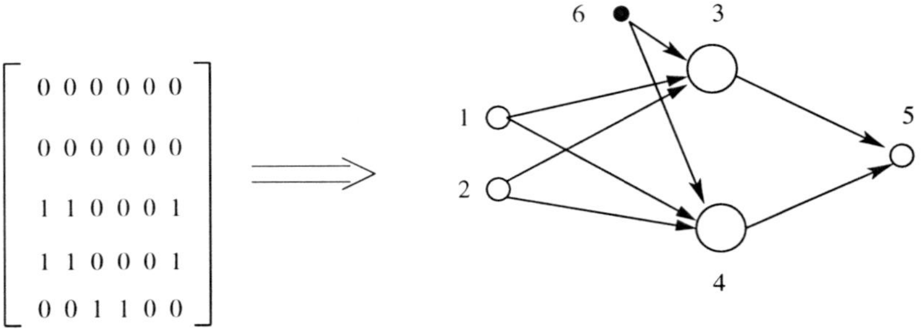

Fig. 4.2. An example of a connection matrix and its corresponding neural network structure

of each connection. Assume that a neural network consists of M neurons in total, including the input and output neurons, then the size of the connection matrix is $M \times (M+1)$, where an element in the last column indicates whether a neuron is connected to a bias value. In the matrix, if element $c_{ij}, i = 1, ..., M, j = 1, ..., M$ equals 1, it means that there is a connection between the i-th and j-th neuron and the signal flows from neuron j to neuron i. If $j = M+1$, it indicates that there is a bias in the i-th neuron. Figure 4.2 illustrates a connection matrix and the corresponding network structure. It can be seen from the figure that the network has two input neurons, two hidden neurons, and one output neuron. Besides, both hidden neurons have a bias.

The strength (weight) of the connections is defined in the weight matrix. Accordingly, if c_{ij} in the connection matrix equals zero, the corresponding element in the weight matrix must be zero too.

4.4.2 Evolutionary Variations and Life-time Learning

A genetic algorithm has been used for optimizing the structure and weights of the neural networks. Binary coding has been used representing the neural network structure and real-valued coding for encoding the weights. Five genetic operations have been introduced in the global search, four of which mutate the connection matrix (neural network structure) and one of which mutates the weights. The four mutation operators are insertion of a hidden neuron, deletion of a hidden neuron, insertion of a connection and deletion of a connection (21). A Gaussian-type mutation is applied to mutate the weight matrix. No crossover has been employed in this algorithm.

After mutation, an improved version of the Rprop algorithm (22) has been employed to train the weights. This can be seen as a kind of life-time learning (the first objective only) within a generation. After learning, the fitness of each individual with regard to the approximation error (f_1) is updated. In addition, the weights modified during the life-time learning are encoded back to the chromosome, which is known as the Lamarckian type of inheritance.

The Rprop learning algorithm (29) is believed to be a fast and robust learning algorithm. Let w_{ij} denotes the weight connecting neuron j and neuron i, then the change of the weight (Δw_{ij}) in each iteration is as follows:

$$\Delta w_{ij}^{(t)} = -\text{sign}\left(\frac{\partial E^{(t)}}{\partial w_{ij}}\right) \cdot \Delta_{ij}^{(t)}, \tag{4.10}$$

where $sign(\cdot)$ is the sign function, $\Delta_{ij}^{(t)} \geq 0$ is the step-size, which is initialized to Δ_0 for all weights. The step-size for each weight is adjusted as follows:

$$\Delta_{ij}^{(t)} = \begin{cases} \xi^{+} \cdot \Delta_{ij}^{(t-1)} &, \text{ if } \frac{\partial E^{(t-1)}}{\partial w_{ij}} \cdot \frac{\partial E^{(t)}}{\partial w_{ij}} > 0 \\ \xi^{-} \cdot \Delta_{ij}^{(t-1)} &, \text{ if } \frac{\partial E^{(t-1)}}{\partial w_{ij}} \cdot \frac{\partial E^{(t)}}{\partial w_{ij}} < 0, \\ \Delta_{ij}^{(t-1)} &, \text{ otherwise} \end{cases} \tag{4.11}$$

where $0 < \xi^{-} < 1 < \xi^{+}$. To prevent the step-sizes from becoming too large or too small, they are bounded by $\Delta_{\text{min}} \leq \Delta_{ij} \leq \Delta_{\text{max}}$.

One exception must be considered. After the weights are updated, it is necessary to check if the partial derivative changes sign, which indicates that the previous step might be too large and thus a minimum has been missed. In this case, the previous weight change should be retracted:

$$\Delta w^{(t)} = -\Delta_{ij}^{(t-1)}, \text{ if } \frac{\partial E^{(t-1)}}{\partial w_{ij}} \cdot \frac{\partial E^{(t)}}{\partial w_{ij}} < 0. \tag{4.12}$$

Recall that if the weight change is retracted in the t-th iteration, the $\partial E^{(t)} / \partial w_{ij}$ should be set to 0.

In reference (22), it is argued that the condition for weight retraction in Equation (4.12) is not always reasonable. The weight change should be retracted only if the partial derivative changes sign and if the approximation error increases. Thus, the weight retraction condition in Equation (4.12) is modified as follows:

$$\Delta w^{(t)} = -\Delta_{ij}^{(t-1)}, \text{ if } \frac{\partial E^{(t-1)}}{\partial w_{ij}} \cdot \frac{\partial E^{(t)}}{\partial w_{ij}} < 0 \textbf{ and } E^{(t)} > E^{(t-1)}. \tag{4.13}$$

It has been shown on several benchmark problems in (22) that the modified Rprop (termed as Rprop^{+} in (22)) exhibits consistently better performance than the Rprop algorithm.

4.4.3 Crowded Tournament Selection

To select the offspring for the next generation, we employ the crowded tournament selection method proposed in the NSGA-II algorithm (17). At first, the offspring and the parent populations are combined. Then, a non-dominated rank and a local crowding distance are assigned to each individual in the combined population. In the non-dominated ranking, the non-dominated solutions are found out and assigned a rank 1. These solutions consist of the first non-dominated front. After that, the non-dominated solutions with rank 1 are removed from the population. Then, non-dominated solutions in the rest of the individuals are identified, which is the second non-dominated front. A rank of 2 is assigned to these solutions. This procedure repeats until all the individuals are assigned to a non-dominated front. In the next step, a crowding distance is calculated for each individual with regard to the non-dominated front it belongs to. The crowding distance of solution i in the non-dominated front j is the distance of the two neighboring of solution s_i^j in the objective space.

$$d_i^j = \sum_{k=1}^{m} |f_k(s_{i-1}^j) - f_k(s_{i+1}^j)|, \tag{4.14}$$

where m is the number of objectives in the multi-objective optimization problem, solutions s_{i-1}^j and s_{i+1}^j are the two neighboring solutions of solution s_i^j. A large distance is assigned to the boundary solutions in each non-dominated front. Here, the larger the crowding distance is, the less crowded around the solution s_i^j.

In selection, two solutions are chosen randomly. The solution with the better (lower) rank wins the tournament. If the two solutions have the same rank, the one with the larger crowding distance wins. If the two solutions with the same rank and the same crowding distance, choose a winner randomly.

4.5 Illustrative Examples

4.5.1 Experimental Setup

To demonstrate that interpretable logic rules can be extracted from the compact Pareto-optimal solutions, the evolutionary multi-objective algorithm is applied to two benchmark problems, namely, the Breast Cancer Diagnosis data and the Iris data.

Although a non-layered neural network can be generated using the coding scheme described in Section 4.3, feedforward networks with one hidden layer will be generated. The maximum number of hidden nodes is set to 10. The hidden neurons are nonlinear and the output neurons are linear. The activation function used for the hidden neurons is as follows,

$$g(z) = \frac{x}{1 + |x|}, \tag{4.15}$$

which is illustrated in Figure 4.3.

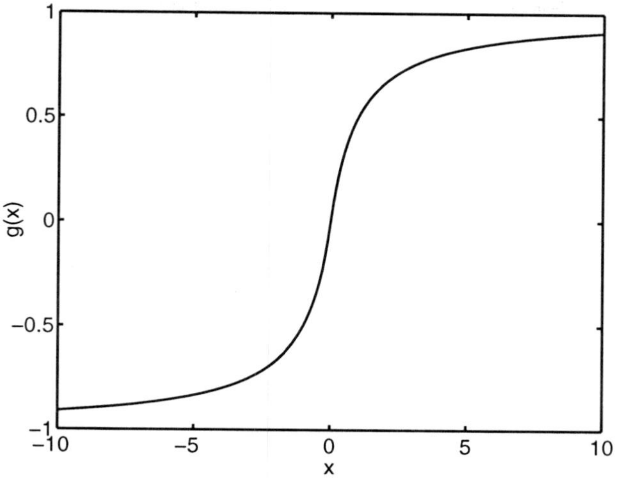

Fig. 4.3. The activation function of the hidden nodes

In this study, the complexity measure defined in Equation (4.6) has been used as the objective describing the complexity of the neural networks.

The population size of the evolutionary algorithm is 100 and the optimization is run for 200 generations. One of the five mutation operations is randomly chosen and performed on each individual. The standard deviation of the Gaussian mutations applied to the weight matrix is set to 0.05. The weights of the network are initialized randomly in the interval of $[-0.2, 0.2]$. In the Rprop$^+$ algorithm, the step-sizes are initialized to 0.0125 and bounded between $[0, 50]$ during the adaptation, and $\xi^- = 0.2$, $\xi^+ = 1.2$, which are the default values recommended in (22) and 50 iterations are implemented in each lifetime learning.

4.5.2 Breast Cancer Diagnosis

The breast cancer benchmark problem in the UCI repository of machine learning database was collected by Dr. W.H. Wolberg at the University of Wisconsin-Madison Hospitals (27). Studies have been carried out to extract symbolic rules from trained neural network using the three-step procedure for rule extraction on this benchmark problem (30; 32). The benchmark problem contains 699 examples, each of which has 9 inputs and 2 outputs. The inputs are: clump thickness (x_1), uniformity of cell size (x_2), uniformity of cell shape (x_3), marginal adhesion (x_4), single epithelial cell size (x_5), bare nuclei (x_6), bland chromatin (x_7), normal nucleoli (x_8), and mitosis (x_9). All inputs are normalized, to be more exact, $x_1, ..., x_9 \in \{0.1,\ 0.2,\ ..., 0.8,\ 0.9,\ 1.0\}$. The two outputs are complementary binary value, i.e., if the first output is 1, which means "benign", then the second output is 0. Otherwise, the first output is 0, which means "malignant", and the second output is 1. Therefore, only the first output is considered in this work. The data samples

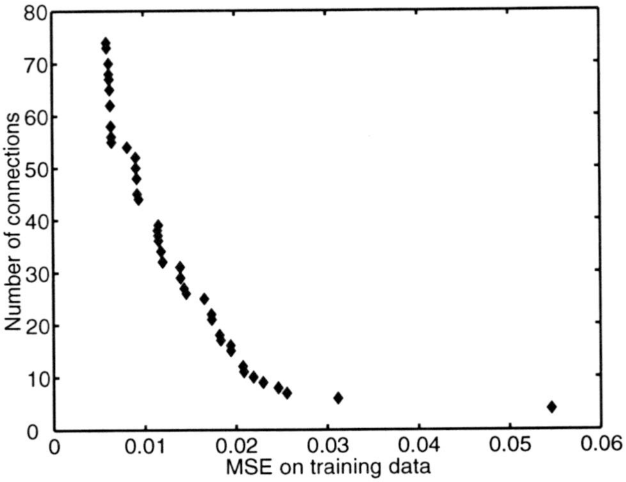

Fig. 4.4. The Pareto front containing 41 non-dominated solutions representing neural networks of a different model complexity

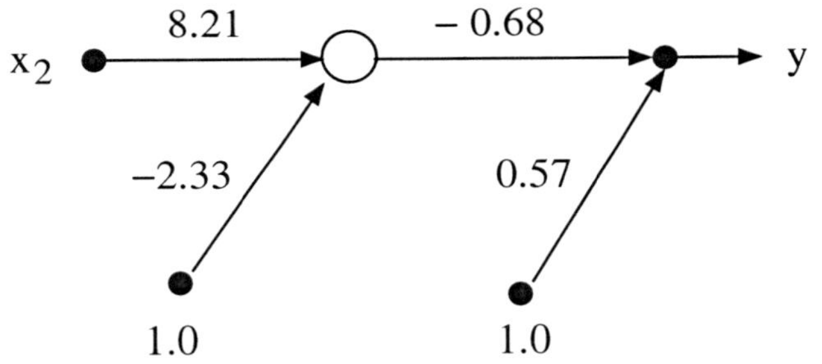

Fig. 4.5. The structure of the simplest Pareto-optimal neural network

are divided into two groups: one training data set containing 599 samples and one test data set containing 100 samples. The test data are unavailable to the algorithm during the evolution.

The non-dominated solutions obtained at the 200-th generation are plotted in Figure 4.4. Note that many solutions in the final population are the same and finally 41 non-dominated solutions have been generated.

Among the 41 neural networks, the simplest one has only 4 connections: 1 input node, one hidden node and 2 biases, see Figure 4.5. The mean squared error (MSE) of the network on the training and test data are 0.0546 and 0.0324, respectively.

Assuming that a case can be decided to be "malignant" if $y < -0.75$, and "benign" if $y > 0.75$. For the neural network in Figure 4.5, if the output of the hidden node is z, we have:

$$-0.68z + 0.57 > 0.75, \tag{4.16}$$

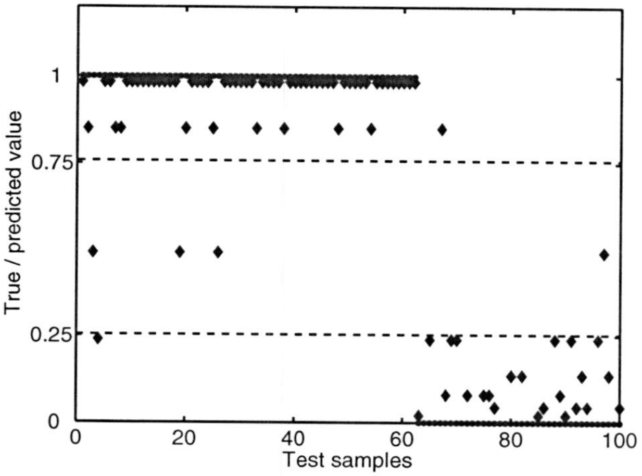

Fig. 4.6. The prediction results of the simplest Pareto-optimal neural network on test data

which means that

$$z < -0.28. \tag{4.17}$$

Let a denote the summed input of the hidden node, we obtain

$$\frac{a}{1+|a|} < -0.28, \tag{4.18}$$

which implies that

$$a < -0.39. \tag{4.19}$$

Therefore, we get:

$$8.21x_2 - 2.33 < -0.39, \tag{4.20}$$

and finally we have:

$$x_2 < 0.21. \tag{4.21}$$

Thus, we can derive that if $x_2 < 0.21$, then the case is "benign". The same procedure can be applied to the malignant case.

As a result, the following two logic rules can be extracted from the simplest Pareto-optimal neural network (denoted as MOO_NN1):

$$\text{R1: If } x_2 \text{ (uniformity) } \geq 0.5\text{, then malignant;} \tag{4.22}$$

$$\text{R2: If } x_2 \text{ (uniformity) } \leq 0.2\text{, then benign.} \tag{4.23}$$

Based on these two simple rules, only 2 out of 100 test samples will be misclassified, and 4 of them cannot be decided with a predicted value of 0.49, which is very ambiguous. The prediction results on the test data are presented in Fig 4.6.

Now let us look at the second simplest Pareto-optimal neural network, which has 6 connections in total. The connection and weights of the network are given in

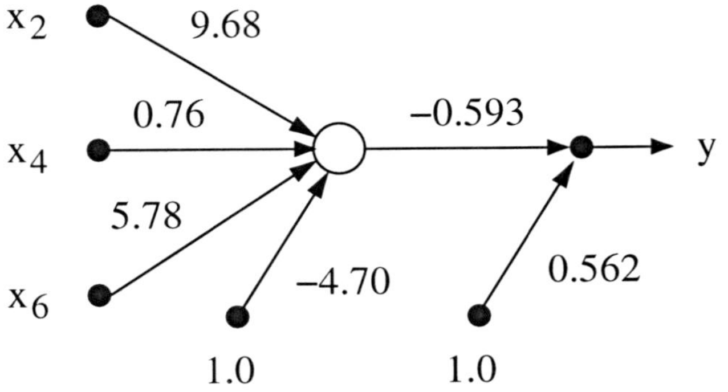

Fig. 4.7. The structure of the second simplest Pareto-optimal neural network

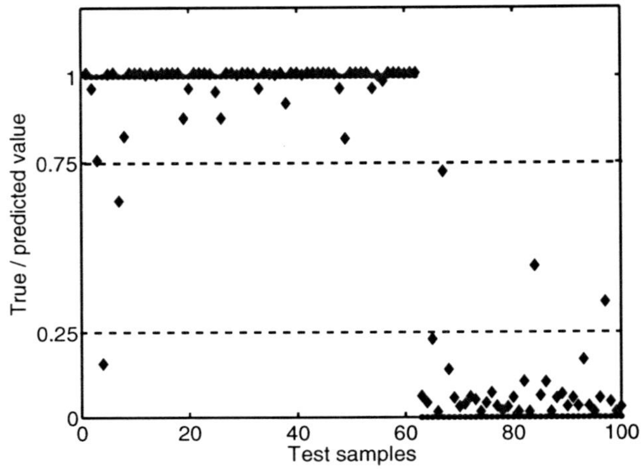

Fig. 4.8. The prediction results of the second simplest Pareto-optimal neural network on test data

Figure 4.7, and the prediction results are provided in Figure 4.8. The MSE of the network on training and test data are 0.0312 and 0.0203, respectively.

In this network, x_2, x_4 and x_6 are present. If the same assumptions are used in deciding whether a case is benign or malignant, then we could extract the following rules: (denoted as MOO_NN2)

$$
\begin{array}{lrl}
\text{R1: If} & x_2 \text{ (uniformity) } \geq 0.6 & \text{or} \\
& x_6 \text{ (bare nuclei) } \geq 0.9 & \text{or} \\
& x_2 \text{ (uniformity) } \geq 0.5 \wedge x_6 \text{(bare nuclei) } \geq 0.2 & \text{or} \\
& x_2 \text{ (uniformity) } \geq 0.4 \wedge x_6 \text{ (bare nuclei) } \geq 0.4 & \text{or} \\
& x_2 \text{ (uniformity) } \geq 0.3 \wedge x_6 \text{ (bare nuclei) } \geq 0.5 & \text{or} \\
& x_2 \text{ (uniformity) } \geq 0.2 \wedge x_6 \text{ (bare nuclei) } \geq 0.7, & \text{then malignant;}
\end{array}
\quad (4.24)
$$

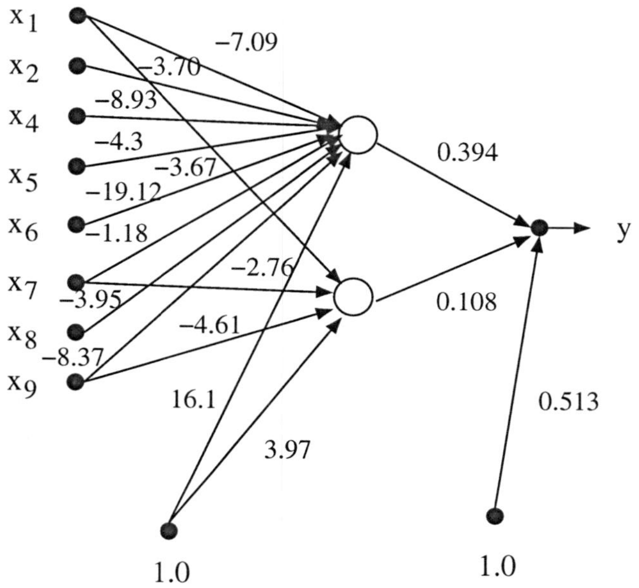

Fig. 4.9. The Pareto-optimal neural network with 16 connections

R2: If x_2 (uniformity) $\leq 0.1 \wedge x_6$ (bare nuclei) ≤ 0.4 or
x_2 (uniformity) $\leq 0.2 \wedge x_6$ (bare nuclei) ≤ 0.2, then benign; (4.25)

Compared to the simplest network, with the introduction of two additional features x_6 and x_4 (although the influence of x_4 is too small to be reflected in the rules), the number of cases that are misclassified has been reduced to 1, whereas the number of cases on which no decision can be made remains to be 4, although the ambiguity of the decision for the four cases do decrease.

The above two neural networks are very simple in structure. We have shown that for such networks of a low model complexity, interpretable logic rules can be extracted. In the following, we will take a look at two neural networks obtained in the multi-objective optimization, which are of better accuracy but are of more signal-type quality, i.e., no interpretable rules can be extracted.

The first network of relatively higher model complexity has 16 connections, whose structure and weights are described in Figure 4.9. The prediction results are plotted in Figure 4.10. In this network, only x_3 is absent and there are 2 hidden nodes. The MSE on training and test data sets are 0.019 and 0.014, respectively. From Figure 4.10, we can see that the classification accuracy is better: only two cases are misclassified. However, extracting symbolic rules from the network becomes much more difficult. Besides, although the architecture of the two simple networks still

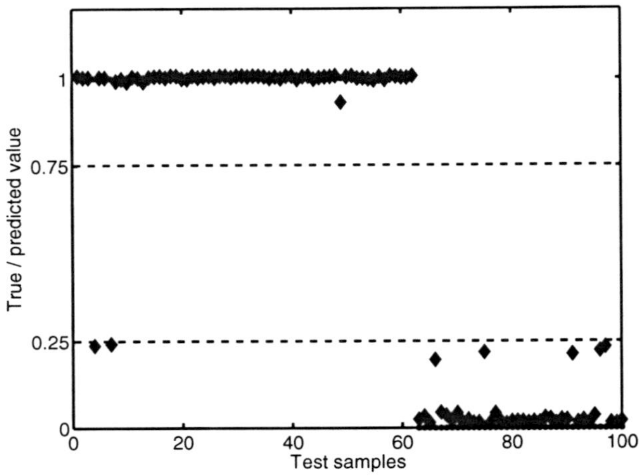

Fig. 4.10. The prediction results of the Pareto-optimal neural network on test data

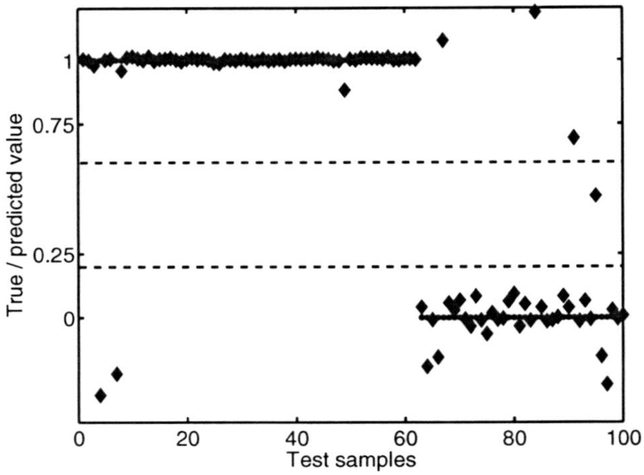

Fig. 4.11. The prediction results of the most complex Pareto-optimal neural network obtained in the simulation

exist in the current network, it not longer shows a dominating influence. Thus, the "skeleton" defined by the simple networks has been lost.

The most complex network obtained in the run has 74 connections. All input features are included in the network and the number of hidden nodes is 9. The MSE on the training data set is 0.0060, however, the MSE on the test data set increases to 0.066 with 5 samples misclassified and 1 undetermined. It seems that the network has over-fitted the training data and the understanding of the network is difficult. The prediction results of this neural network are provided in Figure 4.11.

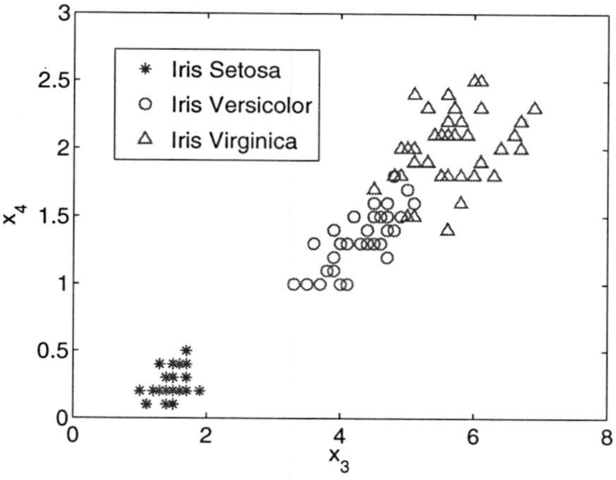

Fig. 4.12. The distribution of the Iris data set

4.5.3 The Iris Data

The second data set we looked at is the Iris data which was originated from references (27; 33). The data set contains 3 classes of 40 instances each, where each class refers to a type of iris plant. The three classes are: Iris Setosa (class 1, represented by -1), Iris Versicolor (class 2, represented by 0), and Iris Virginica (class 3, represented by 1). Four attributes are used to predict the iris class, i.e., sepal length (x_1), sepal width (x_2), petal length (x_3), and petal width (x_4), all in centimeter. Among the three classes, class 1 is linearly separable from the other two classes, and class 2 and 3 are not linearly separable from each other, refer to Figure 4.12.

The same parameter settings have been used to generate the Pareto-optimal neural networks that minimize the accuracy and the complexity. The final population contains 11 non-nominated solutions, which are plotted in Figure 4.13. The most compact Pareto-optimal solution has 8 connections in total, and only attribute x_3 is used for prediction, see Figures 4.14 and 4.15 for the structure of the neural network and the prediction results. Although this simple network is not able to separate all the three classes, we can extract the following single rule, which is able to separate class 1 from the other two classes.

$$\text{R1: If } x_3 \text{ (petal length)} \leq 2.2, \text{ then Iris Setosa.} \tag{4.26}$$

If we look at the class distribution in Figure 4.13, we find that this simple rule effectively captures the condition under which class 1 can be separated from the other 2 classes. Let us now take a closer look at the input-output relation, refer to Figure 4.18, we find that attribute x_3 and x_4 alone are able to separate class 1. However, it is clearly more advantageous to choose attribute x_3 than x_4, and our optimization method has selected the optimal feature for separating class 1 from classes 2 and 3.

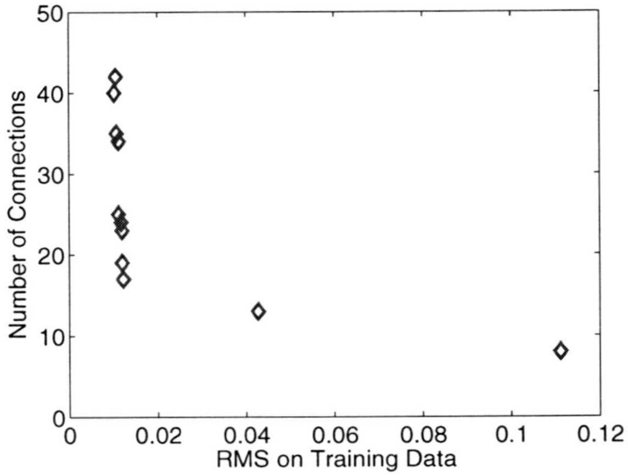

Fig. 4.13. Pareto-optimal solutions generated from the Iris data

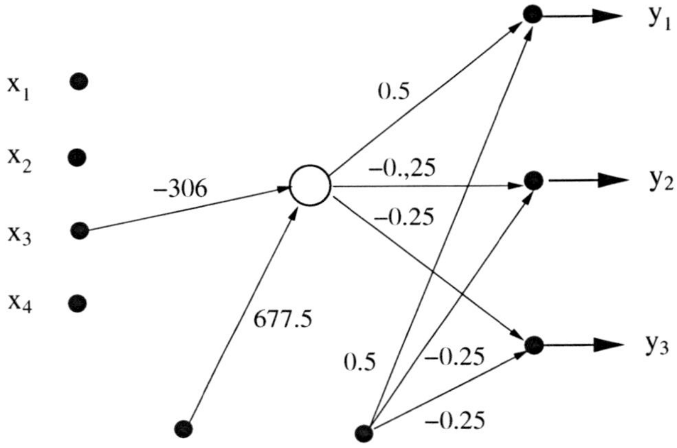

Fig. 4.14. The neural network structure of the simplest Pareto-optimal solution

Now let us investigate the second simplest Pareto-optimal neural network. Similarly, we can extract the following two logic rules:

R1: If x_3 (petal length) $\leq 2.2 \wedge x_4$ (petal width) ≤ 1.0, then Iris Setosa; or

R2: If x_3 (petal length) $> 2.2 \wedge x_4$ (petal width) ≤ 1.4, then Iris Versicolor; or

R3: If x_4 (petal width) ≥ 1.8, then Iris Virginica; (4.27)

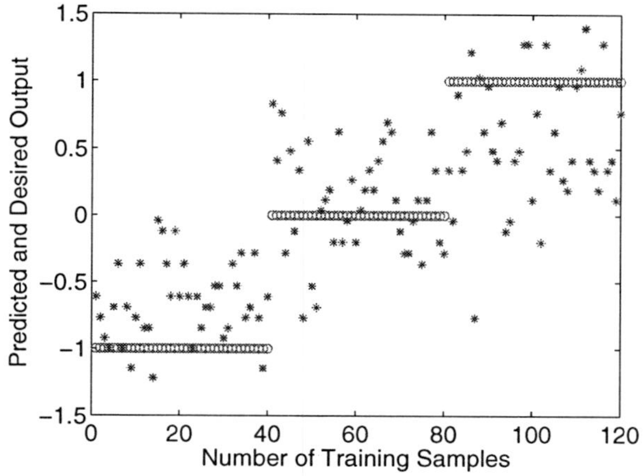

Fig. 4.15. The desired and predicted results of the simplest Pareto-optimal neural network on the training data

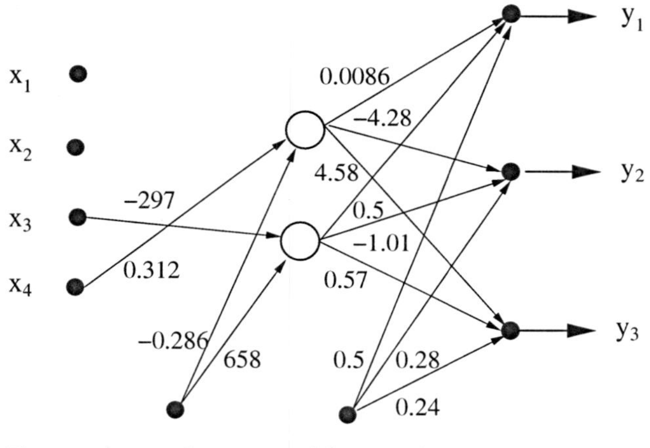

Fig. 4.16. The neural network structure of the second simplest Pareto-optimal solution

The correctness of the rules can be checked by looking at the input-output distribution of the data in Figure 4.18.

4.6 Conclusions

Extracting correct and interpretable rules from data is of great interest for data mining. This chapter suggests a method for generating a set of Pareto-optimal using an evolutionary multi-objective optimization algorithm. Among the Pareto-optimal neural networks, we show that interpretable logic rules can be easily extracted from the

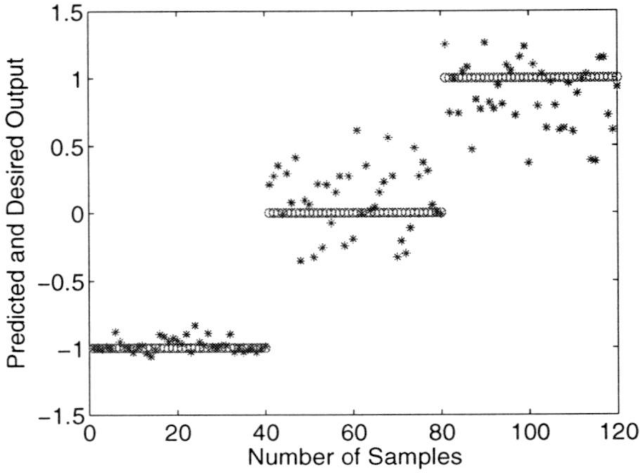

Fig. 4.17. The desired and predicted results of the second simplest Pareto-optimal neural network on the training data

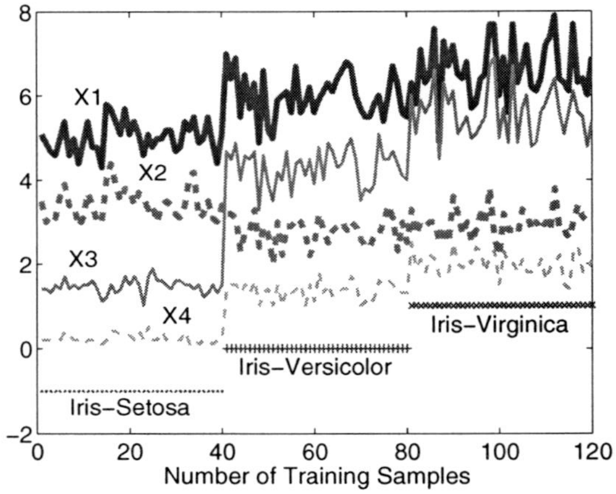

Fig. 4.18. The input-output distribution of the Iris data

compact Pareto-optimal solutions. This idea has been verified on the Breast Cancer Diagnosis data set and the Iris data set. We show that the compact Pareto-optimal neural networks, though very simple and not perfect in terms of approximation accuracy, are able to select the most relevant attributes and effectively separate the classes in the data set. When the complexity increase, it is then difficult to extract understandable logic rules from the neural networks, though the approximation accuracy is better.

As discovered in our previous study (11), the Pareto-optimal solutions that are located in the knee part of the Pareto-front are the most likely the ones that generalize well on unseen data. Together with the findings in this work, we show that Pareto-based multi-objective approach to machine learning is advantageous over traditional approaches where different objective functions are summed up. This advantage is made possible by achieving a Pareto front consisting of a number of solutions, which is able to reveal much deeper insights into the system than a single neural network.

References

[1] Andrew R, Diederich J, Tickle A (1995) A survey and critique of techniques for extracting rules from trained artificial neural networks. Knowledge Based Systems 8(6):373–389

[2] Jin Y, Sendhoff B (2003) Extracting interpretable fuzzy rules from RBF networks. Neural Processing Letters 17(2):149–164

[3] Jin Y (2003) Advanced fuzzy systems design and applications. Springer, Heidelberg

[4] Duch W, Setiono R, Zurada J (2004) Computational intelligence methods for rule-based data understanding. Proceedings of the IEEE 92(5):771–805

[5] Kolman E, Margaliot M (2005) Are artificial neural networks white boxes? IEEE Transactions on Neural Networks 16(4):844–852

[6] Etchell T A, Lisboa P J G (2006) Orthogonal search-based rule extraction (OSRE) for trained neural networks: a practical and efficient approach. IEEE Transactions on Neural Networks 17(2):374–38

[7] Mitra S, Pal S K, Mitra P (2002) Data mining in soft computing framework: A survey. IEEE Transactions on Neural Networks 13(1):3–14

[8] Jin Y, Seelen W V, Sendhoff B (1999) Generating FC3 fuzzy rule systems from data using evolution strategies. IEEE Transactions on Systems, Man, and Cybernetics – Part B: Cybernetics 29(6):829–845

[9] Jin Y (2000) Fuzzy modeling of high-dimensional systems: Complexity reduction and interpretability improvement. *IEEE Transactions on Fuzzy Systems* 8(2):212–221

[10] Jin Y, Sendhoff B, Körner E (2005) Evolutionary multi-objective optimization for simultaneous generation of signal-type and symbol-type representations. The Third International Conference on Evolutionary Multi-Objective Optimization, 752–766

[11] Jin Y, Sendhoff B, Körner E (2006) Simultaneous generation of accurate and interpretable neural network classifiers. In Y. Jin (eds) Multi-objective Machine Learning, 291–312

[12] Jin Y (2006) (eds) Multi-objective machine learning. Springer, Berlin.

[13] Bishop C M (1995) Neural networks for pattern recognition. Oxford University Press, Oxford, UK

[14] Burnham K P, Anderson D R (2002) Model selection and multimodel inference. Springer, New York

[15] Coello Coello C A, Veldhuizen D, Lamont G (2002) Evolutionary algorithms for solving multi-objective problems. Kluwer Academic, New York

[16] Deb K (2001) Multi-objective optimization using evolutionary algorithms. Wiley, Chichester

[17] Deb K, Agrawal S, Pratap A, Meyarivan T (2000) A fast elitist non-dominated sorting genetic algorithm for multi-objective optimization: NSGA-II. In Parallel Problem Solving from Nature 6:849–858

[18] Duch W, Adamczak R, Grabczewski K (1998) Extraction of logical rules from backpropagation networks. Neural Processing Letters 7:1–9

[19] Fodor J A, Pylyshyn Z W (1998) Connectionism and cognitive architecture: A critical analysis. Cognition 28(3):3–71

[20] Gabrieli J, Poldrack R, Desmond J (1998) The role of left prefrontal cortex in langrange and memory. Proceedings of the National Academy of Sciences 95:906–913

[21] Hüsken M, Gayko J E, Sendhoff B (2000) Optimization for problem classes–Neural networks that learn to learn. In Yao X, Fogel D B (eds) IEEE Symposium on Combinations of Evolutionary Computation and Neural Networks (ECNN 2000), IEEE Press, 98–109

[22] Igel C, Hüsken M (2000) Improving the Rprop learning algorithm. In Proceedings of the 2nd ICSC International Symposium on Neural Computation, 115–121

[23] Ishibuchi H, Yamamoto T (2003) Evolutionary multiobjective optimization for generating an ensemble of fuzzy rule-based classifiers. In Proceedings of the Genetic and Evolutionary Computation Conference, 1077–188

[24] Ishikawa M (2000) Rule extraction by successive regularization. Neural Networks 13:1171–1183

[25] Jin Y, Okabe T, Sendhoff B (2004) Neural network regularization and ensembling using multi-objective evolutionary algorithms. In Congress on Evolutionary Computation, IEEE 1–8

[26] Miller D A, Zurada J M (1998) A dynamical system perspective of structural learning with forgetting. IEEE Transactions on Neural Networks 9(3):508–515

[27] Prechelt L (1994) PROBEN1 - a set of neural network benchmark problems and benchmarking rules. Technical Report 21/94, Fakultät für Informatik, Universität Karlsruhe

[28] Reed R D, Marks II R J (1999) Neural smithing. The MIT Press

[29] Riedmiller M, Braun H (1993) A direct adaptive method for faster backpropagation learning: The RPROP algorithm. In IEEE International Conference on Neural Networks, 586–591

[30] Setiono R (2000) Generating concise and accurate classification rules for breast cancer diagnosis. Artificial Intelligence in Medicine 18:205–219

[31] Setiono R, Liu H (1996) Symbolic representation of neural networks. IEEE Computer 29(3):71–77

[32] Taha I, Ghosh J (1999) Symbolic interpretation of artificial neural networks. IEEE Transactions on Knowledge and Data Engineering 11(3):448–463

[33] Anderson E (1935) The rises of the gaspe peninsula. Bulletin of the American Iris Society 59:2–5

5

On the Usefulness of MOEAs for Getting Compact FRBSs Under Parameter Tuning and Rule Selection*

R. Alcalá, J. Alcalá-Fdez, M.J. Gacto, F. Herrera

Department of Computer Science and Artificial Intelligence
University of Granada, E-18071 – Granada, Spain
{alcala, jalcala, herrera}@decsai.ugr.es, mjgacto@ugr.es

Summary. In the last years, multi-objective genetic algorithms have been successfully applied to obtain Fuzzy Rule-Based Systems satisfying different objectives, usually different performance measures.

Recently, multi-objective genetic algorithms have been also applied to improve the difficult trade-off between interpretability and accuracy of Fuzzy Rule-Based Systems, obtaining linguistic models not only accurate but also interpretable. It is know that both requirements are usually contradictory, however, a multi-objective genetic algorithm can obtain a set of solutions with different degrees of accuracy and interpretability.

This contribution briefly reviews the state of the art in this very recent topic and presents an approach in order to prove the ability of multi-objective genetic algorithms for getting compact fuzzy rule-based systems under rule selection and parameter tuning, i.e., to obtain linguistic models with improved accuracy and the least number of possible rules. This way to work involves another trade-off degree respect with the works in the existing literature which has been still not explored.

5.1 Introduction

One of the most important areas for the application of Fuzzy Set Theory are Fuzzy Rule-Based Systems (FRBSs). Typically, they have been considered to solve problems in different application domains as classification, regression or control and rule mining (26). There are at least two different kinds of FRBSs in the literature, Mamdani (33) and Takagi-Sugeno (36), which differ on the composition of the rule consequent. The use of one or another depends on the fact that the main requirement is the interpretability or the accuracy of the model, respectively.

Many automatic techniques have been proposed in the literature to extract a proper set of fuzzy rules from numerical data. However, most of these techniques usually try to improve the performance associated to the prediction error without

* Supported by the Spanish CICYT Project TIN2005-08386-C05-01 (KEEL II).

R. Alcalá et al.: *On the Usefulness of MOEAs for Getting Compact FRBSs Under Parameter Tuning and Rule Selection*, Studies in Computational Intelligence (SCI) **98**, 91–107 (2008)
www.springerlink.com

inclusion of any interpretability measure, an essential aspect of FRBSs. In the last years, the problem of finding the right trade-off between interpretability and accuracy, in spite of the original nature of fuzzy logic, has arisen a growing interest in methods which take both aspects into account (1). Of course, the ideal thing would be to satisfy both criteria to a high degree, but since they are contradictory issues generally it is not possible.

Recently, Multi-Objective Evolutionary Algorithms (MOEAs) (5; 12) have been also applied to improve the difficult trade-off between interpretability and accuracy of FRBSs, obtaining linguistic models not only accurate but also interpretable. Since this problem presents a multi-objective nature the use of these kinds of algorithms, to obtain a set of solutions with different degrees of accuracy and interpretability, is an interesting way to work. In this way, we can obtain a set of solutions where each solution tends to satisfy a criterion to a higher extent than another, and considering different performance and interpretability measures an expert can select those solutions satisfying these objectives (accuracy and interpretability measures) to the desired degree.

Most of these works apply MOEAs to obtain accurate but also interpretable Mamdani FRBSs (8; 19; 22; 23; 24; 25; 27; 32) since they are much more interpretable than Takagi-Sugeno ones (29; 30; 38; 39). This contribution briefly reviews the state of the art in this recent topic, analyzing the most representative works of the specialized literature in order to point out the most important aspects that should been taken into account to deal with these kinds of problems. All these works try to obtain the complete Pareto (set of non-dominated solutions with different trade-off) by selecting or learning the set of rules better representing the example data, i.e., improving the system accuracy and decreasing the FRBS complexity but not considering learning or tuning of the membership function parameters. In this way, this work also presents an approach in order to study the usefulness of multi-objective genetic algorithms to obtain simpler and still accurate linguistic fuzzy models by applying rule selection and a tuning of membership functions, which represents a more complex search space and therefore needs of different considerations respect to the works in the existing literature. Two different MOEAs are considered and studied to perform this task.

This contribution is arranged as follows. The next section analyzes the state of the art on the use of MOEAs to get a better trade-off between interpretability and accuracy of FRBSs in different application areas. In Section 5.3, we present two different algorithms by considering two of the most known MOEAs in order to obtain linguistic models by applying rule selection together with a tuning of parameters. Section 5.4 shows an experimental study of these methods applied to a complex but interesting problem. Finally, Section 5.5 points out some conclusions and further research lines.

5.2 Use of MOEAs to Get the Interpretability-accuracy Trade-off of Fuzzy Systems

Evolutionary algorithms deal simultaneously with a set of possible solutions (the so-called population) which allows us to find several members of the Pareto optimal set in a single run of the algorithm. Additionally, evolutionary algorithms are less susceptible to the shape or continuity of the Pareto front (e.g., they can easily deal with discontinuous and concave Pareto fronts) (5; 6; 12; 41). In this way, MOEAs have proved to be very effective to search for optimal solutions to problems that incorporate multiple performance criteria in competition with each other (41; 42). These kinds of algorithms generate a family of equally valid solutions, where each solution tends to satisfy a criterion to a higher extent than another.

After the first generation of MOEAs for general use (17; 18; 35) (MOGA, NPGA and NSGA respectively) a second generation of MOEAs (3; 4; 9; 10; 11; 13; 14; 31; 40; 41; 43) started when elitism became a standard mechanism (34). The Strength Pareto Evolutionary Algorithm 2 (SPEA2) (43) and the Nondominated Sorting Genetic Algorithm II (NSGA-II) (11; 13) can be considered as the most representative MOEAs of the second generation. However, nowadays NSGA-II is the paradigm within the MOEA research community since the powerful crowding operator that this algorithm incorporates usually allows to obtain the widest Pareto sets in a great variety of problems, and this is a very appreciated property in this framework.

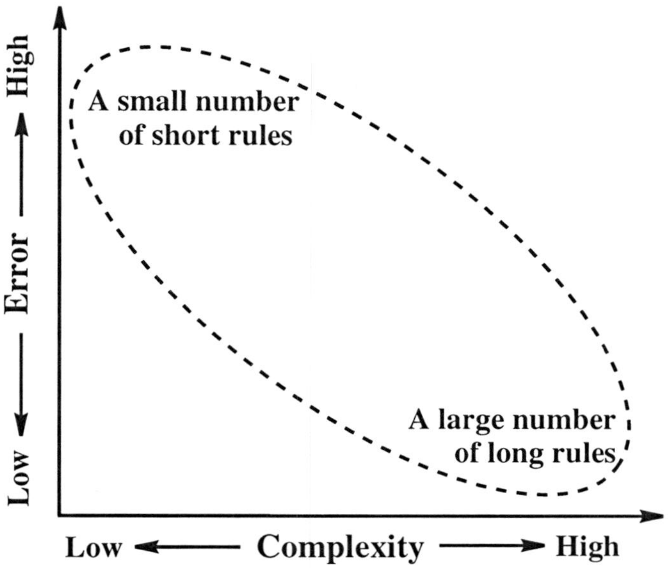

Fig. 5.1. Trade-off between the error and the interpretability of rule sets

MOEAs have been also applied to improve the difficult trade-off between interpretability and accuracy of FRBSs, where each solution in the pareto front represents a different trade-off between interpretability and accuracy (see Figure 5.1). The most continuous and prolific research activity in the application of MOEAs to Mamdani FRBS generation for finding the accuracy-interpretability trade off has been certainly performed by Ishibuchi's group. Since they suggested the idea of a MOEA to find a set of solutions with different trade off by using rule selection and two objectives (19), they have published many works in this interesting field. Earlier works (19; 21; 22) were devoted to the application of simple MOEAs of the first generation to perform a rule selection on an initial set of classification rules involving *"don't care"* conditions and considering two different objectives (classification accuracy and number of rules). Then, a third objective was also included in order to minimize the length of the rules by rule selection (23; 24) or rule learning (23). In (27), they apply a better MOEA, the Multi-Objective Genetic Local Search (20) (MOGLS), following the same approach for rule selection with three objectives. And finally, two algorithms based on a MOEA of the second generation (NSGA-II) have been proposed respectively for rule selection (32) and rule learning (28) considering the same concepts. In the literature, we can also find some papers of other researchers in this topic. For instance in (8), Cordon et al. use MOGA for jointly performing feature selection and fuzzy set granularity learning with only two objectives.

At this point, we can see that all the methods mentioned were applied to classification problems for rule selection or rule learning, without learning or tuning the membership functions that were initially fixed. Most of the works in this topic only consider quantitative measures of the system complexity in order to improve the interpretability of such systems, rarely considering qualitative measures. Moreover, MOEAs considered were slight modifications of MOEAs proposed for general use (MOGA, NSGA-II, etc.) or based on them. Notice that, although NSGA-II improves the results respect to other MOEAs, since to cross non-dominated rule sets with very different numbers of rules and different rule structures (forced by the NSGA-II crowding operator) usually gives a bad accuracy, this MOGA needed of an adaptation to favor the cross of similar solutions in order to also get good results for the accuracy objective (32). The problem is that, although to directly apply this powerful algorithm to this problem gets a wider Pareto front with several good solutions, to improve the accuracy objective is more difficult than simplifying the fuzzy models, by which the Pareto front finally obtained still becomes sub-optimal respect to the accuracy objective.

On the other hand, there are a few works in the framework of fuzzy modeling. In (25), authors show how a simple MOGA can be applied to a three-objective optimization problem (again not considering learning or tuning of parameters). Some applications of MOEAs have been also discussed in the literature to improve the difficult trade-off between accuracy and interpretability of Takagi-Sugeno models. In (29; 30; 38; 39), accuracy, interpretability and compactness have been considered as objectives to obtain interpretable and very accurate Takagi-Sugeno models. However, since Takagi-Sugeno models have a linear function in the consequent part of each fuzzy rule, they are close to accuracy representing another type of trade-off with less interpretable models (25). For this reason, the type of rule most used to achieve the trade-off between accuracy and complexity are the fuzzy rules with linguistic terms in both the antecedent and consequent parts, i.e., Mamdani rules (33).

Table 5.1 presents a resume of the different methods in the existing literature.

Table 5.1. Use of MOEAs for finding the accuracy-interpretability trade off

Year	Ref.	Problem	MOEA/Gen.	#Objs.	RS	FS	RL	LP
MAMDANI LINGUISTIC MODELS (closer to interpretability)								
1995/7/8	(19; 21; 22)	Classification	MOGA/1st	2	√	√	-	-
2001	(8)	Classification	MOGA/1st	2	-	√	√	-
2001	(23; 24)	Classification	MOGA/1st	3	√	√	√	-
2003	(25)	Regression	MOGA/1st	3	√	√	√	-
2004	(27)	Classification	MOGLS/1st	3	√	√	-	-
2005	(32)	Classification	NSGA-II*/2nd	3	√	√	-	-
2007	(28)	Classification	NSGA-II*/2nd	3	-	√	√	-
TAKAGI-SUGENO MODELS (closer to accuracy)								
2001	(29; 30)	Regression	Specific/1st	3	-	-	√	√
2005	(38)	Regression	MOGA*/1st	5	√	√	√	√
2005	(39)	Regression	NSGA-II*/2nd	5	√	√	√	√

RS = Rule Selection, **FS** = Feature Selection, **RL** = Rule Learning, **LP** = Learning/Tuning of parameters.
* based on that algorithm

5.3 Two Different MOEAs for Rule Selection and Tuning of Parameters

As we explain in the previous section most works in the field of fuzzy systems are applied to classification problems by learning or selecting rules, not considering tuning of the membership function parameters. The main reason of this fact is that a tuning of parameters implies a lost of the interpretability to some degree. However, it is known that this way to work greatly improves the performance of the linguistic models so obtained, being another alternative to improve the interpretability-accuracy trade-off. For this reason, we present two different algorithms that focus the research in the linguistic fuzzy modeling area, in order to evaluate the performance of MOEAs in a field which is still less explored, and with the main objective of inject some ideas or recommendations in this open topic (improvement of the interpretability of very accurate models).

The proposed algorithms will perform rule selection from a given fuzzy rule set together with the parametric tuning of the membership functions. To do that, we apply the most used multi-objective algorithms of the second generation, SPEA2 (43)

and NSGA-II (13), considering two different objectives, system error and number of rules.

In the next subsections, we present the SPEA2 and NSGA-II algorithms applied for linguistic fuzzy modeling. At first, the common components of both algorithms are proposed and then the main steps and characteristic of both algorithms are described.

5.3.1 Main Components of the Algorithms

As mentioned, we use the well-known SPEA2 and NSGA-II to perform rule selection and tuning of membership functions and with the aim of improving the desired trade-off between interpretability and accuracy. In the following, the components needed to apply these algorithms in this concrete problem are explained. They are coding scheme, initial gene pool, objectives and genetic operators:

- **Coding scheme and initial gene pool**
  A double coding scheme for both *rule selection* (C_S) and *tuning* (C_T) is used:

$$C^p = C_S^p C_T^p$$

  In the $C_S^p = (c_{S1}, \ldots, c_{Sm})$ part, the coding scheme consists of binary-coded strings with size m (with m being the number of initial rules). Depending on whether a rule is selected or not, values '1' or '0' are respectively assigned to the corresponding gene. In the C_T part, a real coding is considered, being m^i the number of labels of each of the n variables comprising the database,

$$\begin{array}{l} C_i = (a_1^i, b_1^i, c_1^i, \ldots, a_{m^i}^i, b_{m^i}^i, c_{m^i}^i), \quad i = 1, \ldots, n, \\ C_T^p = C_1 C_2 \ldots C_n \ . \end{array}$$

  The initial population is obtained with all individuals having all genes with value '1' in the C_S part. And in the C_T part the initial database is included as first individual. The remaining individuals are generated at random within the corresponding variation intervals. Such intervals are calculated from the initial database. For each membership function $C_i^j = (a^j, b^j, c^j)$, the variation intervals are calculated in the following way:

$$\begin{array}{r} [I_{a^j}^l, I_{a^j}^r] = [a^j - (b^j - a^j)/2, a^j + (b^j - a^j)/2] \\ [I_{b^j}^l, I_{b^j}^r] = [b^j - (b^j - a^j)/2, b^j + (c^j - b^j)/2] \\ [I_{c^j}^l, I_{c^j}^r] = [c^j - (c^j - b^j)/2, c^j + (c^j - b^j)/2] \end{array}$$

- **Objectives**
  Two objectives are minimized for this problem: the number of rules (interpretability) and the Mean Squared Error (accuracy),

$$\text{MSE} = \frac{1}{2 \cdot |E|} \sum_{l=1}^{|E|} (F(x^l) - y^l)^2,$$

with $|E|$ being the size of a data set E, $F(x^l)$ being the output obtained from the FRBS decoded from the said chromosome when the l-th example is considered and y^l being the known desired output. The fuzzy inference system considered to obtain $F(x^l)$ is the *center of gravity weighted by the matching* strategy as defuzzification operator and the *minimum t-norm* as implication and conjunctive operators.

- **Genetic Operators**
  The crossover operator depends on the chromosome part where it is applied: the BLX-0.5 (16) in the C_T part and the HUX (15) in the C_S part:
  - In the C_T part, the BLX-0.5 crossover is used. This operator is based on the the concept of environments (the offspring are generated around one parent). These kinds of operators present a good cooperation when they are introduced within evolutionary models forcing the convergence by pressure on the offspring.
    The BLX operator can be described as follows. To ease the description let us assume that $X = (x_1 \cdots x_n)$ and $Y = (y_1 \cdots y_n)$, $(x_i, y_i \in [a_i, b_i] \subset \Re, i = 1 \cdots n)$, are two real-coded chromosomes that are going to be crossed. The BLX operator (with $\alpha = 0.5$) generates one descendent $Z = (z_1, \cdots, z_g)$ where z_i is a randomly (uniformly) chosen number from the interval $[l_i, u_i]$, with $l_i = max\{a_i, c_{min} - I\}$, $u_i = min\{b_i, c_{max} + I\}$, $c_{min} = min\{x_i, y_i\}$, $c_{max} = max\{x_i, y_i\}$ and $I = (c_{max} - c_{min}) \cdot \alpha$.
  - In the C_S part, the HUX crossover is used. The HUX crossover exactly interchanges the mid of the alleles that are different in the parents (the genes to be crossed are randomly selected among those that are different in the parents). This operator ensures the maximum distance of the offspring to their parents (exploration).

  Finally, four offspring are generated by combining the two from the C_S part with the two from the C_T part (the two best replace to their parent). Moreover, the mutation operator changes a gene value at random in the C_S and C_T parts (one in each part) with probability P_m.

5.3.2 SPEA2 Based Approach

The SPEA2 algorithm (43) was designed to overcome the problems of its predecessor, the SPEA algorithm (41). In contrast with SPEA, SPEA2: (1) incorporates a fine-grained fitness assignment strategy which takes into account for each individual the number of individuals that it dominates and the number of individuals by which it is dominated; (2) uses the nearest neighbour density estimation technique which guides the search more efficiently; (3) has an enhanced archive truncation method which guarantees the preservation of boundary solutions. Next, we briefly describe the complete SPEA2 algorithm.

SPEA2 uses a fixed population and archive size. The population forms the current base of possible solutions, while the archive contains the current solutions. The archive is constructed and updated by copying all non-dominated individuals in both archive and population into a temporary archive. If the size of this temporary archive differs from the desired archive size, individuals are either removed or added as necessary. Individuals are added by selecting the best dominated individuals, while the removal process uses a heuristic clustering routine in the objective space. The motivation for this is that one would like to try to ensure that the archive contents represent distinct parts of the objective space. Finally, when selecting individuals for participating in the next generation all candidates are selected from the archive using a binary tournament selection scheme.

Considering the components defined and the descriptions of the authors in (43), the SPEA2 algorithm consists of the next steps:

Input: N (population size),
$\overline{N}$ (external population size),
T (maximum number of generations).
Output: A (non-dominated set).

1. Generate an initial population P_0 and create the empty external population $\overline{P}_0 = \emptyset$.
2. Calculate fitness values of individuals in P_t and $\overline{P}_t$.
3. Copy all non-dominated individuals in $P_t \cup \overline{P}_t$ to $\overline{P}_{t+1}$. If $|\overline{P}_{t+1}| > \overline{N}$ apply truncation operator. If $|\overline{P}_{t+1}| < \overline{N}$ fill with dominated in $P_t \cup \overline{P}_t$.
4. If $t \geq T$, return A and stop.
5. Perform binary tournament selection with replacement on $\overline{P}_{t+1}$ in order to fill the mating pool.
6. Apply recombination (BLX-HUX) and mutation operators to the mating pool and set P_{t+1} to the resulting population. Go to step 2 with $t = t + 1$.

5.3.3 NSGA-II Based Approach

The NSGA-II algorithm (11; 13) is one of the most well-known and frequently-used MOEAs in the literature. As in other evolutionary algorithms, first NSGA-II generates an initial population. Then an offspring population is generated from the current population by selection, crossover and mutation. The next population is constructed from the current and offspring populations. The generation of an offspring population and the construction of the next population are iterated until a stopping condition is satisfied. The NSGA-II algorithm has two features, which make it a high-performance MOEA. One is the fitness evaluation of each solution based on Pareto ranking and a crowding measure, and the other is an elitist generation update procedure.

Each solution in the current population is evaluated in the following manner. First, Rank 1 is assigned to all non-dominated solutions in the current population. All solutions with Rank 1 are tentatively removed from the current population. Next, Rank 2 is assigned to all non-dominated solutions in the reduced current population. All solutions with Rank 2 are tentatively removed from the reduced current population. This procedure is iterated until all solutions are tentatively removed from the current population (i.e., until ranks are assigned to all solutions). As a result, a different rank is assigned to each solution. Solutions with smaller ranks are viewed as being better than those with larger ranks. Among solutions with the same rank, an additional criterion called a crowding measure is taken into account.

The crowding measure for a solution calculates the distance between its adjacent solutions with the same rank in the objective space. Less crowded solutions with larger values of the crowding measure are viewed as being better than more crowded solutions with smaller values of the crowding measure.

A pair of parent solutions are selected from the current population by binary tournament selection based on the Pareto ranking and the crowding measure. When the next population is to be constructed, the current and offspring populations are combined into a merged population. Each solution in the merged population is evaluated in the same manner as in the selection phase of parent solutions using the Pareto ranking and the crowding measure. The next population is constructed by choosing a specified number (i.e., population size) of the best solutions from the merged population. Elitism is implemented in NSGA-II algorithm in this manner.

Considering the components previously defined and the descriptions of the authors in (13), NSGA-II consists of the next steps:

1. A combined population R_t is formed with the initial parent population P_t and offspring population Q_t (initially empty).
2. Generate all non-dominated fronts $F = (F_1, F_2, ...)$ of R_t.
3. Initialize $P_{t+1} = 0$ and $i = 1$.
4. Repeat until the parent population is filled.
5. Calculate crowding-distance in F_i.
6. Include i-th non-dominated front in the parent population.
7. Check the next front for inclusion.
8. Sort in descending order using crowded-comparison operator.
9. Choose the first $(N - |P_{t+1}|)$ elements of F_i.
10. Use selection, crossover (BLX-HUX) and mutation to create a new population Q_{t+1}.
11. Increment the generation counter.

5.4 Experiments: A Case Study on Linguistic Fuzzy Modeling

In this section, we present an example on the use of MOGAs to obtain linguistic models with a good trade-off between interpretability and accuracy in a real-world problem (7) with 4 input variables that consists of estimating the maintenance costs of medium voltage lines in a town. The methods considered for the experiments are briefly described in Table 5.2.

WM method is considered to obtain the initial rule base to be tuned. T and S methods perform the tuning of parameters and rule selection respectively. TS indicates tuning together with rule selection in the same algorithm. All of them consider the accuracy of the model as the sole objective. The proposed algorithms (NSGA-II and SPEA2) perform rule selection from a given fuzzy rule set together with the

Table 5.2. Methods considered for comparison

Méthod	**Ref.**	**Description**
WM	(37)	Wang & Mendel algorithm
WM+T	(2)	Tuning of Parameters
WM+S	(2)	Rule Selection
WM+TS	(2)	Tuning and Rule Selection
SPEA2	-	Tuning and Rule Selection with SPEA2
NSGA-II	-	Tuning and Rule Selection with NSGA-II

parametric tuning of the membership functions considering two objectives, system error and number of rules.

In the next subsections, we describe this real-world problem and finally we show the results obtained.

5.4.1 Problem Description and Experiments

Estimating the maintenance costs of the medium voltage electrical network in a town (7) is a complex but interesting problem. Since a direct measure is very difficult to obtain, it is useful to consider models. These estimations allow electrical companies to justify their expenses. Moreover, the model must be able to explain how a specific value is computed for a certain town. Our objective will be to relate the *maintenance costs of the medium voltage lines* with the following four variables: *sum of the lengths of all streets in the town*, *total area of the town*, *area that is occupied by buildings*, and *energy supply to the town*. We will deal with estimations of minimum maintenance costs based on a model of the optimal electrical network for a town in a sample of 1,059 towns.

To develop the different experiments, we consider a *5-folder cross-validation model*, i.e., 5 random partitions of data each with 20%, and the combination of 4 of them (80%) as training and the remaining one as test. For each one of the 5 data partitions, the tuning methods have been run 6 times, showing for each problem the averaged results of a total of 30 runs. In the case of methods with multi-objective approach (SPEA2 and NSGA-II), the averaged values are calculated considering the most accurate solution from each Pareto obtained. In this way, the multi-objective algorithms are compared with several single objective based methods. This way to work differs with the previous works in the specialized literature in which one or several Pareto fronts are presented and an expert should after select one solution. Our main aim following this approach is to compare the same algorithm by only considering an accuracy objective (WM+TS) with the most accurate solution found by the multi-objective ones in order to see if the Pareto fronts obtained are not only wide but also optimal (similar solutions to that obtained by WM+TS should be included in the final Pareto).

The initial linguistic partitions are comprised by *five linguistic terms* with equally distributed triangular shape membership functions. Figure 5.2 shows an example of this type of partition.

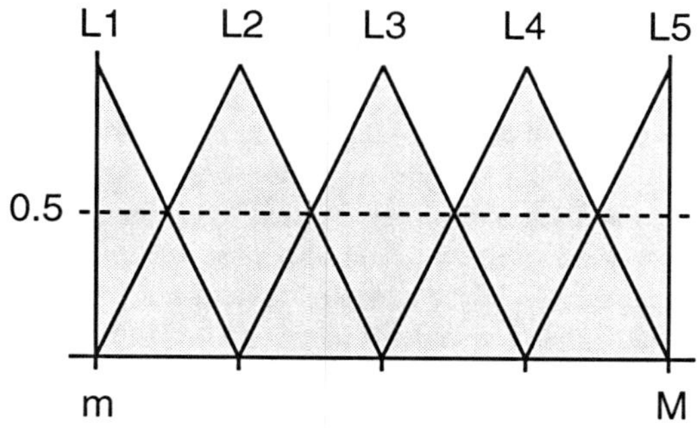

Fig. 5.2. Example of linguistic partition with five linguistic terms

The values of the input parameters considered by MOEAs are shown in the next: population size of 200, external population size of 61 (in the case of SPEA2), 50000 evaluations and 0.2 as mutation probability per chromosome.

5.4.2 Results and Analysis

The results obtained by the analyzed methods are shown in Table 5.3, where $\#R$ stands for the number of rules, MSE_{tra} and MSE_{tst} respectively for the averaged error obtained over the training and test data, σ for the standard deviation and *t-test* for the results of applying a *test t-student* (with 95 percent confidence) in order to ascertain whether differences in the performance of the multi-objective approach are significant when compared with that of the other algorithms in the table. The interpretation of this column is:

- $\star$ represents the best averaged result.
- $+$ means that the best result has better performance than that of the corresponding row.

Table 5.3. Results obtained by the studied methods

Method	#R	MSE_{tra}	σ_{tra}	t-test	MSE_{tst}	σ_{tst}	t-test
WM	65	57605	2841	+	57934	4733	+
WM+T	65	18602	1211	+	22666	3386	+
WM+S	40.8	41086	1322	+	59942	4931	+
WM+TS	41.9	14987	391	+	18973	3772	=
SPEA2	**33**	**13272**	1265	*	**17533**	3226	*
NSGA-II	41.0	14488	965	=	18419	3054	=

Analyzing the results showed in Table 5.3 we can highlight the two following facts:

- NSGA-II obtains the same accuracy and the same number of rules than the models obtained with WM+TS (single objective-based approach) considering the most accurate result of each obtained Pareto. Therefore, we could consider that this algorithm gets good solutions, from the most accurate ones (with more complexity) to the most simple ones (with the worst accuracy).
- The SPEA2 method shows a reduction of the MSE_{tra} and produces more or less the same MSE_{tst} respect to the models obtained by only considering the accuracy objective (WM+TS). Moreover, a considerable number of rules have been removed from the initial FRBS, obtaining simpler models with a similar performance. In this way, the most accurate models obtained by SPEA2 considering a multi-objective approach get a better trade-off between interpretability and accuracy than those obtained by a single objective based algorithm (which theoretically should obtain the most accurate results).

These results are due to the large search space that involves this problem. There are some initial rules that should be removed since they do not cooperate in a good way with the remaining ones. Even in the case of only considering an accuracy-based objective, the large search space that supposes the tuning of parameters makes very difficult to remove these kinds of rules since bad rules are tuned together with the remaining ones searching for their best cooperation. The use of a multi-objective approach favors a better selection of the ideal number of rules, preserving some rule configurations until the rule parameters are evolved to dominate solutions including bad rules.

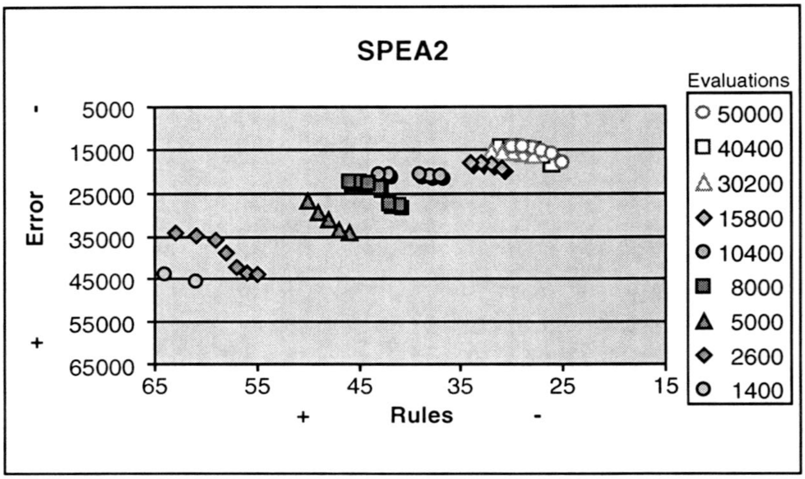

Fig. 5.3. Evolution of the Pareto fronts of SPEA2

On the other hand, NSGA-II tries to obtain a wider Pareto front by crossing very different solutions based on its crowding operator. However, in this problem, it is difficult to obtain accurate solutions by favoring the crossing of solutions with very different rule configurations (those in the Pareto), which try to obtain the optimum by learning very different parameters for the membership functions. This is the reason by which this algorithm does not work as well as SPEA2 in this particular problem.

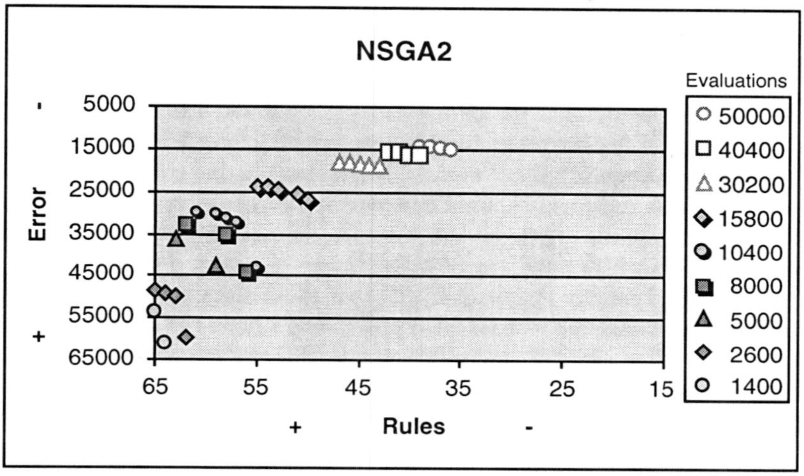

Fig. 5.4. Evolution of the Pareto fronts of NSGA2

All these facts suggest the design of more specific algorithms in order to get even better solutions for these kinds of problems, probably, solutions with a better performance that considering a single objective and with a minor number of rules. In Figures 5.3 and 5.4, we can see the Pareto evolution for each multi-objective algorithm. We can observe as the Pareto moves along without having a wide extension, even in the case of NSGA-II.

5.5 Concluding Remarks

In this work we have analyzed the use of MOEAs to improve the trade-off between interpretability and accuracy of FRBSs. A brief revision of the state of the art in this topic has been performed. From this study we can point out the following facts:

- Most of the contributions in this topic were made in the framework of fuzzy classification, considering Mamdani FRBSs.
- Most of the works only consider quantitative measures of the system complexity to determine the FRBS interpretability.
- None of the works considered a learning or tuning of the membership functions, only performing a rule learning or selection.

- The MOEAs considered were slight modifications of MOEAs proposed for general use (MOGA, NSGA-II, etc.) or adaptations of them for this concrete and difficult problem. It is due to the special nature of this problem, in which to improve the accuracy objective is more difficult than simplifying the fuzzy models, by which the Pareto front finally obtained still becomes sub-optimal respect to the accuracy objective. This specially occurs in algorithms as NSGA-II (32), since to cross non-dominated rule sets with very different numbers of rules and different rule structures (forced by the NSGA-II crowding operator) usually gives a bad accuracy, by which this MOEA needs of an adaptation to favor the cross of similar solutions in order to also get good results for the accuracy objective (28; 32).

On the other hand, this contribution has presented two different algorithms and a case of study on the use of multi-objective genetic algorithms to obtain simpler and still accurate linguistic fuzzy models by also considering a tuning of the system parameters, which represents a more complex search space and therefore needs of different considerations respect to the works in the existing literature. The results obtained have shown that the use of MOEAs can represent a way to obtain even more accurate and simpler linguistic models than those obtained by only considering performance measures. In this case (also performing a tuning of the parameters), the problem of crossing very different solutions with different number of rules and very different parameters becomes more important since to obtain a wide Pareto with the best solutions is practically impossible. Therefore, as further work, more specific algorithms should be proposed in order to get the best possible solutions.

References

[1] Casillas J, Cordón O, Herrera F, Magdalena L (2003) (eds), Interpretability issues in fuzzy modeling. Studies in Fuzziness and Soft Computing 129, Springer, Heidelberg, Germany

[2] Casillas J, Cordón O, del Jesus M J, Herrera F (2005) Genetic tuning of fuzzy rule deep structures preserving interpretability and its interaction with fuzzy rule set reduction. IEEE Transactions on Fuzzy Systems 13(1):13–29

[3] Coello Coello C A, Toscano G (2001) A micro-genetic algorithm for multiobjective optimization. First International Conference on Evolutionary Multi-Criterion Optimization, Lecture Notes in Computer Science 1993:126–140

[4] Coello Coello C A, Toscano G (2001) Multiobjective optimization using a micro-genetic algorithm. Proceedings of the Genetic and Evolutionary Computation Conference (San Francisco, California), 274–282

[5] Coello Coello C A, Van Veldhuizen D A, Lamont G B (2002) Evolutionary algorithms for solving multi-objective problems. Kluwer Academic Publishers

[6] Coello Coello C A (2006) Evolutionary multiobjective optimization: A historical view of the field. IEEE Computational Intelligence Magazine 1(1):28–36

[7] Cordón O, Herrera F, Sánchez L (1999), Solving electrical distribution problems using hybrid evolutionary data analysis techniques. Applied Intelligence 10:5–24

[8] Cordon O, Herrera F, del Jesus M J, Villar P (2001) A multiobjective genetic algorithm for feature selection and granularity learning in fuzzy-rule based classification systems. Proceedings of IX IFSA World Congress and XX NAFIPS International Conference, Vancouver, Canada, 1253–1258
[9] Corne D, Knowles J, Oates M (2000) The pareto envelope-based selection algorithm for multiobjective optimization. Proceedings of the Parallel Problem Solving from Nature VI Conference, Lecture Notes in Computer Science 1917:839–848
[10] Corne D, Jerram N, Knowles J, Oates M (2001) PESA-II: Region based selection in evolutionary multiobjective optimization. Proceedings of the Genetic and Evolutionary Computation Conference, San Francisco, Califormia, 283–290
[11] Deb K, Agrawal S, Pratab A, Meyarivan T (2000) A fast elitist non-dominated sorting genetic algorithm for multi-objective optimization: NSGA-II. Proceedings of the Parallel Problem Solving from Nature VI Conference, Lecture Notes in Computer Science 1917, Paris, France, 849–858
[12] Deb K (2001) Multi-objective optimization using evolutionary algorithms, John Wiley & Sons, Chichester, UK
[13] Deb K, Agrawal S, Pratab A, Meyarivan T (2002) A fast and elitist multiobjective genetic algorithm: NSGA-II. IEEE Transactions on Evolutionary Computation 6(2):182-197
[14] Erickson M, Mayer A, Horn J (2001) The niched Pareto genetic algorithm 2 applied to the design of groundwater remediation systems. Proceedings of the First International Conference on Evolutionary Multi-Criterion Optimization, Lecture Notes in Computer Science 1993, London, UK, 681–695
[15] Eshelman L J (1991) The CHC adaptive search algorithm: How to have safe search when engaging in nontraditional genetic recombination. Foundations of genetic Algorithms 1:265–283
[16] Eshelman L J, Schaffer J D (1993) Real-coded genetic algorithms and interval-schemata. Foundations of Genetic Algorithms 2:187–202
[17] Fonseca C M, Fleming P J (1993) Genetic algorithms for multiobjective optimization: Formulation, discussion and generalization. Proceedings of the Fifth International Conference on Genetic Algorithms, San Mateo, California, 416–423
[18] Horn J, Nafpliotis N, Goldberg D E (1994) A niched pareto genetic algorithm for multiobjective optimization. Proceedings of the First IEEE Conference on Evolutionary Computation, IEEE World Congress on Computational Intelligence 1:82–87
[19] Ishibuchi H, Murata T, Turksen I B (1995) Selecting linguistic classification rules by two-objective genetic algorithms. Proceedings of IEEE-SMC'95, Vancouver, Canada, 1410–1415
[20] Ishibuchi H, Murata T (1996) Multi-objective genetic local search algorithm. Proceedings of Third IEEE International Conference Evolutionary Computation, Japan, 119-124

[21] Ishibuchi H, Murata T, Turksen I B (1997) Single-objective and two-objective genetic algorithms for selecting linguistic rules for pattern classification problems. Fuzzy Sets and Systems 89(2):135–150
[22] Ishibuchi H, Murata T, Gen M (1998) Performance evaluation of fuzzy rule-based classification systems obtained by multi-objective genetic algorithms. Computers & Industrial Engineering 35(3-4):575–578
[23] Ishibuchi H, Nakashima T, Murata T (2001) Three-objective genetics-based machine learning for linguistic rule extraction. Information Sciences 136:109–133
[24] Ishibuchi H, Nakashima T, Murata T (2001) Multiobjective optimization in linguistic rule extraction from numerical data. Proceedings of the First International Conference on Evolutionary Multi-Criterion Optimization, Lecture Notes in Computer Science 1993:588–602
[25] Ishibuchi H, Yamamoto T (2003) Interpretability issues in fuzzy genetics-based machine learning for linguistic modelling. In Lawry J, Shanahan J G, Ralescu A L (eds) Modelling with Words: Learning, Fusion, and Reasoning within a Formal Liguistic Representation Framework, Lecture Notes in Computer Science 2873, Springer-Verlag, Berlin, Heidelberg, 209–228
[26] Ishibuchi H, Nakashima T, Nii M (2004) (eds) Classification and modeling with linguistic information granules. Advanced Approaches to Linguistic Data Mining Series: Advanced Information Processing, Springer, Berlin
[27] Ishibuchi H, Yamamoto T (2004) Fuzzy rule selection by multi-objective genetic local search algorithms and rule evaluation measures in data mining. Fuzzy Sets and Systems 141(1):59–88
[28] Ishibuchi H, Nojima Y (2007) Analysis of interpretability-accuracy tradeoff of fuzzy systems by multiobjective fuzzy genetics-based machine learning. International Journal of Approximate Reasoning 44(1):4–31
[29] Jimenez F, Gomez-Skarmeta A F, Roubos H, Babuska R (2001) Accurate, transparent, and compact fuzzy models for function approximation and dynamic modeling through multi-objective evolutionary optimization. Proceedings of First International Conference on Evolutionary Multi-Criterion Optimization, Lecture Notes in Computer Science, 1993:653–667
[30] Jimenez F, Gomez-Skarmeta A F, Roubos H, Babuska R (2001) A multi-objective evolutionary algorithm for fuzzy modelling. Proceedings of 9th IFSA World Congress and 20th NAFIPS International Conference, Vancouver, Canada, 1222–1228
[31] Knowles J D, Corne D W (2000) Approximating the non dominated front using the Pareto archived evolution strategy. Evolutionary Computation 8(2):149–172
[32] Narukawa K, Nojima Y, Ishibuchi H (2005) Modification of evolutionary multiobjectiveoptimization algorithms for multiobjective design of fuzzy rule-based classification systems. Proceedings of 2005 IEEE International Conference on Fuzzy Systems, 809–814
[33] Mamdani E H, Assilian S (1975) An experiment in linguistic synthesis with a fuzzy logic controller. International Journal of Man-Machine Studies 7:1–13

[34] Rudolph G, Agapie A (2000) Convergence properties of some multi-objective evolutionary algorithms. Proceedings of the 2000 Conference on Evolutionary Computation, 1010–1016

[35] Srinivas N, Deb K (1994) Multiobjective optimization using nondominated sorting in genetic algorithms. Evolutionary Computation 2:221-248

[36] Takagi T, Sugeno M (1985) Fuzzy identification of systems and its application to modeling and control. IEEE Transactions on Systems, Man, and Cybernetics 15(1):116–132

[37] Wang L X, Mendel J M (1992) Generating fuzzy rules by learning from examples. IEEE Transactions on Systems, Man, and Cybernetics 22(6):1414–1427

[38] Wang H L, Kwong S, Jin Y C, Wei W, Man K F (2005) Multi-objective hierarchical genetic algorithm for interpretable fuzzy rule-based knowledge extraction. Fuzzy Sets and Systems 149(1):149–186

[39] Wang H L, Kwong S, Jin Y C, Wei W, Man K F (2005) Agentbased evolutionary approach for interpretable rule-based knowledge extraction. IEEE Trans. on Systems, Man, and Cybernetics - Part C: Applications and Reviews 35(2):143–155

[40] Zitzler E, Thiele L (1998) Multiobjective optimization using evolutionary algorithms–a comparative study. In Eiben A (eds) Parallel Problem Solving from Nature V, Springer-Verlag, Amsterdam, 292–301

[41] Zitzler E, Thiele L (1999) Multiobjective evolutionary algorithms: A comparative case study and the strength Pareto approach. IEEE Transactions on Evolutionary Computation 3(4):257–271

[42] Zitzler E, Deb K, Thiele L (2000) Comparison of multiobjective evolutionary algorithms: Empirical results. Evolutionary Computation 8(2):173–195

[43] Zitzler E, Laumanns M, Thiele L (2001) SPEA2: Improving the strength Pareto evolutionary algorithm for multiobjective optimization. Proceedings of the Evolutionary Methods for Design, Optimization and Control with Application to Industrial Problems, Barcelona, Spain, 95-100

6

Classification and Survival Analysis Using Multi-objective Evolutionary Algorithms

Christian Setzkorn

National Center for Zoonosis Research, University of Liverpool, UK
christian@setzkorn.eu

Summary. Model extraction from data has at least three objectives. It aims to produce accurate, comprehensible and interesting models. Hence, multi-objective evolutionary algorithms are a natural choice to tackle this problem. They are capable of optimizing several incommensurable objectives in as single run without making any assumptions about the importance of each objective. This chapter proposes several multi-objective evolutionary algorithms to tackle three different model extraction tasks. The first approach performs supervised classification whilst overcoming some of the shortcomings of existing approaches. The second and third approach tackle two survival analysis problems. All approaches are evaluated on artificial, benchmark and real-world medical data sets.

6.1 Introduction

Multi-Objective Evolutionary Algorithms (MOEAs) are modern heuristic search methods capable of optimizing several incommensurable objectives in a single run without making any assumptions about the importance of each objective. This chapter details three MOEAs to tackle different data mining problems. The first approach is used to induce fuzzy classification rule systems and the other two are used for survival analysis problems. Data mining is a multi-objective task that aims to extract accurate, comprehensible and interesting models from data (21). However, this is not the only reason for the utilization of MOEAs. For example, they can also produce several solutions/models in a single run. Hence, if the preference of the decision maker changes the search does not have to be repeated. If the decision maker prefers more accurate classifiers over more comprehensible classifiers, (s)he could simply choose another solution from the solution set. This helps to save computational resources. There is another reason for our choice of MOEAs. They can be used to prevent overfitting, which can be a serious problem in data mining because overfitted models do not generalize beyond the training data. Many researchers believe that model complexity is one reason for overfitting. This view is also adopted by the author who uses MOEAs to minimize the complexity and increase the accuracy

C. Setzkorn: *Classification and Survival Analysis Using Multi-objective Evolutionary Algorithms*, Studies in Computational Intelligence (SCI) **98**, 109–135 (2008)
www.springerlink.com

of models. The interested reader is referred to (63) for a more detailed discussion of overfitting and alternative methods to prevent it. Several shortcomings of existing approaches were addressed by the implemented MOEAs . For example, many evolutionary approaches have many parameters. In addition, many evolutionary approaches still use accuracy to measure the fit of the model to the data. This is inappropriate when the misclassifications costs and class priors are unknown, which is often the case in practice (24; 28). Hence, we use a measure called the area under the receiver-operating characteristic curve (AUC) (28; 55; 56) that does not have these problems. We also deploy a self-adaptation mechanism to reduce the number of free parameters and we use state of the art multi-objective evolutionary algorithm components, such as a new archiving strategy and the fitness assignment proposed by Laumanns *et al.* (30) and Zitzler *et al.* (39). In addition, we utilized tailor-made representation schemes to maintain the comprehensibility of the models and allow the application of problem-specific variation operators. Another problem of evolutionary algorithms, their high computational demands, has been addressed in an earlier study. It uses the JavaSpaces technology[1] which is free of charge, simplifies the implementation of parallel/distributed applications, and is independent of the underlying hardware and software environment. It can therefore be used to harvest the computational resources of a heterogeneous multi-user computer network in a simple and affordable manner (see also (31)).

This chapter is organized as follows. Section 6.2 details the MOEA for the induction of fuzzy classification rules. Section 6.3.1 details the MOEAs for two survival analysis problems. The chapter concludes in Section 6.4. It has to be noted that this chapter can only provide a brief overview of the implemented MOEAs and related subjects. The interested reader is referred to these publications for more details (22; 50; 52; 63).

6.2 On the Use of Multi-Objective Evolutionary Algorithms for Classification

Supervised classification involves the determination of classifiers from data that were sampled from a particular population. Data consist of vectors (samples) that describe objects, each belonging to exactly one class from a predefined set of classes. Vectors consist of feature values (e.g. age of a patient). The induced classifiers aim to model the process that generated the data by mapping a given object to the correct class (8; 9; 47). This study uses a particular type of classifier known as Fuzzy Classification Rule Systems (FCRS), which is fitted using a MOEA. Rule system usually consists of rules of the following form:

$$IF\ cond_1\ AND\ cond_2\ \ldots\ AND\ cond_n\ THEN\ Class\ =\ \omega_i.$$

[1] JavaSpaces is a trademark of SUN Microsystems, Inc.

The antecedent (IF-part) consists of conjunctions of conditions. Conditions (e.g. $cond_1$) are sets defined upon the domain of features of the data set (e.g. blood pressure = low). Conjunctions correspond to the logical operator AND and combine different sets (e.g. blood pressure = low AND smoking = yes). A consequent (THEN-part) associates a class with an antecedent. As a rule system can contain several antecedents with the same consequent, it can approximating multi-modal distributions, which might be problematic for other models (1; 3; 42). Antecedents with the same consequent are usually combined using disjunctions (logical operator OR) (42).

The majority of rule systems deploy conditions that represent classical sets, which exhibit crisp decision thresholds. An example of a classical set is the set of patients with low blood pressure, whose diastolic blood pressure is below/equal 70 mmHg. Unfortunately, systems that use crisp decision thresholds tend to be unstable as they can produce very different responses (classifications) to similar samples (objects) (35; 58). For example, someone who has a blood pressure of 71 mmHg would not be assigned to the group 'low blood pressure', although her/his blood pressure is only slightly higher than monographsz6 mmHg.

The use of fuzzy sets instead of classical sets can alleviate this problem (59). Fuzzy sets are defined by membership functions that assign values between zero and one to each domain value permitting smooth decision thresholds (16; 59), making this classifier more immune to noise (44; 45) and more suited to model ambiguities (6).

As fuzzy sets produce values between zero and one, other conjunction and disjunction operators have to be defined. The Min and $Prod$ operators are often used to replace AND operator, and the Max operator is used instead of the OR operator. Combined fuzzy sets define high-dimensional prototypical clusters within the feature space whose boundaries are not necessarily axis parallel (36). This is a further advantage of FCRSs and enables them to approximate non-axis parallel distributions (63) using fewer antecedents.

6.2.1 Existing Work

The induction of rule systems with evolutionary algorithms has a long history. The first approaches were proposed in the early 1980s (4). However, most approaches for the induction of fuzzy rule systems have tackled control and regression problems (e.g. (46)). One of the earliest proposal was proposed by Karr *et al.* in 1989 (33). Multi-Objective evolutionary algorithms have only recently been used to induce FCRSs. One usually distinguishes between four approaches for the induction of fuzzy rule systems from data: Michigan approach, Pittsburgh approach, iterative rule learning approach, cooperative co-evolutionary approach. The Michigan approach (4; 26) is the oldest approach and is sometimes referred to as learning classifier systems (23; 27). The Michigan approach was followed by the Pittsburgh approach (64) and the iterative rule learning approach. Recently several hybrids of these approaches have been proposed, such as cooperative co-evolutionary approaches. The interested reader is referred to (63) for an extensive review of existing approaches for the induction of rule systems.

6.2.2 The Implemented Multi-Objective Evolutionary Algorithm

Figure 6.1 depicts the structure of the implemented MOEA, which is similar to that of other evolutionary algorithms (e.g. (7)).

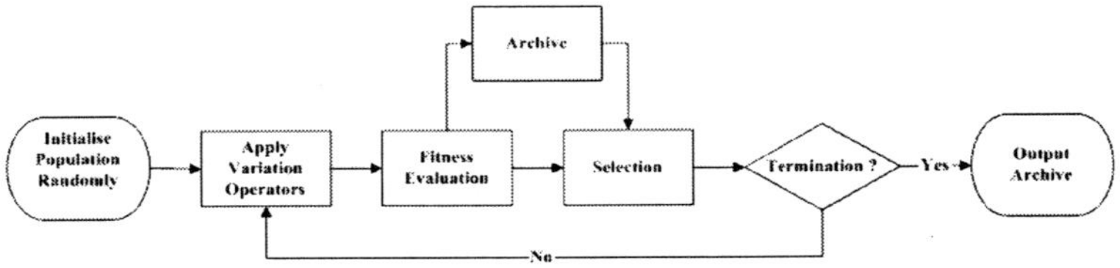

Fig. 6.1. Structure of the implemented MOEA

The system works as follows. First, a number of candidate solutions (FCRSs) are initialized randomly. After this, the variation operators are applied to some of the FCRSs to recombine and/or change them. The fitness evaluation determines each individual's performance (fitness/objective values).

The selection process generates a new population by sampling from the current population and the archive. The archive stores the best (elite) individuals found by the MOEA. This prevents the loss of good solutions (39). The use of an archive is a form of elitism, which can help to produce better solutions (17; 62). In fact, the implemented MOEA uses a new archive, which ensures population diversity and convergence (30). The selection process is followed by the termination test, which either terminates the MOEA or applies the variation operators, repeating the steps. It is expected that better individuals will be produced over time. If the MOEA terminates (e.g. after a maximum number of generations) the FCRSs within the current population and the archive are evaluated on a validation data set that was not used during the induction process. The output of the system is the updated set of individuals in the archive. It follows a more detailed description of each component.

The Representation Scheme

Figure 6.2 depicts an individual's structure. The first part (*Bit Strings*) is used for the self-adaptation. The second part (*Rules*) consists of n rules and confidence values $(CF_1 \ldots CF_n)$ that measures the past performance of a rule (see Equation 6.4).

The number of rules can vary but is restricted by an upper bound. Antecedents $(A_1 \ldots A_n)$ take the form of trees as shown in Figure 6.3. Consequents $(C_1 \ldots C_n)$ are integers that represent the classes.

The antecedent structure is similar to that of genetic programming (GP) (37; 38), with the difference that the antecedents are not combined in one tree. The trees are kept separate to simplify the application of problem-specific variation operators and maintain the comprehensibility of the model. Non-terminal nodes (circles in

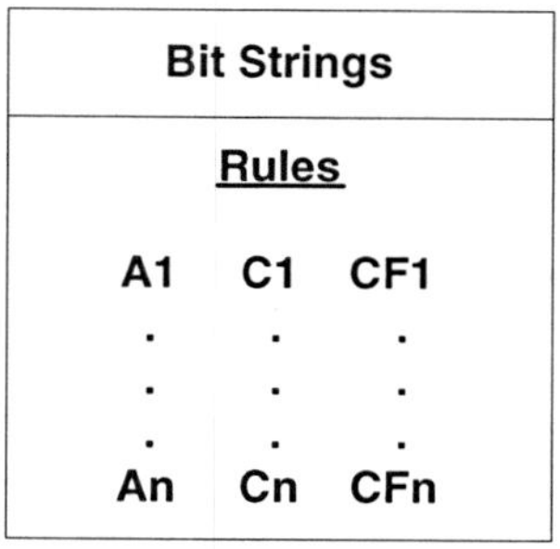

Fig. 6.2. Representation scheme

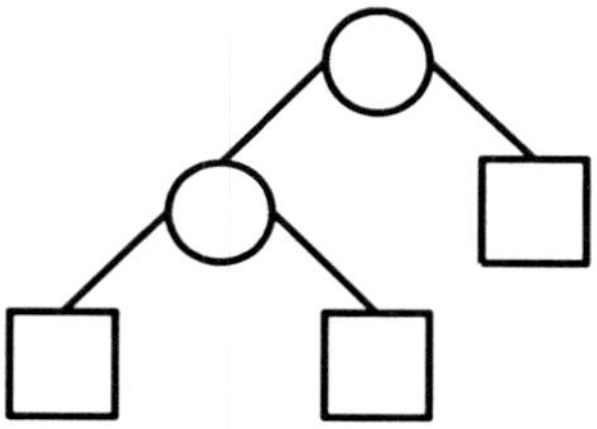

Fig. 6.3. An antecedent tree

Figure 6.3) represent conjunction (Min or the $Prod$ operator). The terminal nodes (squares in Figure 6.3) can be one of the fuzzy sets from Table 6.1.

Table 6.1. Membership functions ($MF1$ - triangular membership function, $MF2$ - trapezoidal membership function, $MF3$ - Gaussian membership function, $MF4$ - bell-shaped membership function, $MF5$ - singleton membership function)

Equation	**Shape**
$MF1(x;a,b,c) = max(min(\frac{x-a}{b-a}, \frac{c-x}{c-b}), 0)$	
$MF2(x;a,b,c,d) = max(min(\frac{x-a}{b-a}, 1, \frac{d-x}{d-c}), 0)$	
$MF3(x;a,b) = e^{\frac{-(x-b)^2}{2a^2}}$	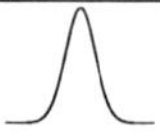
$MF4(x;a,b,c) = \frac{1}{1+\lvert\frac{x-c}{a}\rvert^{2b}}$	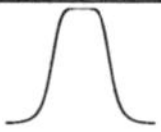
$MF5(x;a,b) = \begin{cases} a & \text{if } x = b \\ 0 & \text{otherwise} \end{cases}$	

If the feature is numerical or ordinal $MF1 \dots MF4$ are used. If it is nominal the singleton membership function ($MF5$) is used, as the domain values are unordered. There are further restrictions to make the trees more comprehensible. Features can only be used once and the size of a tree, although adaptable, is limited.

Antecedent trees generate values between zero and one as response to samples. If a FCRS contains several antecedents for the same class their maximum response value is chosen as the output value for this class. If the FCRS does not contain antecedents for a class its response is zero. Hence, the individual produces a vector with membership degrees for each class.

It is important to note that this MOEA is a hybrid between the Pittsburgh and Michigan approach as models can contain several rules for different classes (Pittsburgh approach) or one or several rules for one class (Michigan approach).

Fitness Evaluation

The MOEA utilises the fitness assignment of SPEA2 (39), which uses the Pareto dominance relation (see Equation 6.7) and population density information to prevent premature convergence. We adopt the view that the Pareto dominance relation is the only basis on which a solution can be said to perform better than another in the total absence of information about the importance of the objectives (25). The MOEA optimizes three objectives; o_1 measures the fit of the FCRS to the data whereas o_2 and o_3 measure the complexity of the FCRS. Objective o_1, is computed according to Equation 6.1 (54) and has to be minimized.

$$o_1 = 1 - (\frac{2}{c(c-1)} \sum_{i<j} A(i,j)). \tag{6.1}$$

Here c denotes the number of classes to be predicted and $A(i,j)$ is computed according to Equation 6.2.

$$A(i,j) = \frac{A(i \mid j) + A(j \mid i)}{2}. \tag{6.2}$$

Both $A(i \mid j)$ and $A(j \mid i)$ estimate the AUC using the Mann-Whitney-Wilcoxon (MWW) two-sample test statistic. This statistic compares two one-dimensional arrays. The first array contains the maximum responses of antecedents with the consequent (class) indicated by the first index (i in $A(i \mid j)$ or j in $A(j \mid i)$) to objects that belong to this class. The second array contains the maximum responses from antecedents with the consequent (class) that is indicated by the second index (j in $A(i \mid j)$ or i in $A(j \mid i)$). If the antecedents discriminate well between objects from two classes, the first array should contain larger values than the second array. This can be measured using the MWW statistic, which is determined by merging the two arrays and arranging them in ascending order (without losing the information of whether an array value originated from the first or the second array). After this, Equation 6.3 is applied.

$$A = \frac{S_0 - n_0(n_0 + 1)/2}{n_0 n_1}. \tag{6.3}$$

Here S_0 denotes the sum of the ranks of values within the first array. The values n_0 and n_1 denote the number of values in the first and second array respectively. The certainty degree CF measures a rule's past performance (57) and is computed according to Equation 6.4.

$$CF = \frac{Sj^k}{S^k}. \tag{6.4}$$

Here Sj^k denotes the sum of the response values from the k-th rule's antecedent to samples that belong to the class indicated by its consequent and S^k denotes the response values of the k-th rule's antecedent to any sample (regardless of its class membership). The number of rules (o_2) and features (o_3) within the FCRS is used to measure its comprehensibility or complexity. Both objectives have to be minimized. The actual fitness $F(i)$ of an individual i is computed according to Equation 6.5.

$$F(i) = R(i) + D(i). \tag{6.5}$$

Here $R(i)$ captures dominance information (see Equation 6.6 and 6.7) and $D(i)$ captures density information (see Equation 6.8) of the i-th individual.

$$R(i) = \sum_{j \in P_t + \overline{P_t}, j \succ i} S(j) \tag{6.6}$$

$$S(i) = \mid \{ j \mid j \in P_t + \overline{P_t}. \wedge i \succ j \} \mid . \tag{6.7}$$

Here P_t and $\overline{P_t}$ refer to individuals from the population and the archive respectively. The expression $i \succ j$ denotes the dominance relation between individual i and j. An individual i dominates another individual j if i is at least as good as j in all objectives and better with respect to at least one objective. Equation 6.6 determines the number of individuals within P_t and $\overline{P_t}$ that are dominated by the i-th individual and the number of individuals that are dominated by the individuals that dominate the i-th individual. If the value of R_i is zero the individual i is non-dominated. The density information is computed according to Equation 6.7 and is an adaptation of the k-th nearest neighbor method (20).

$$D(i) = \frac{1}{\sigma_i^k + 2}. \tag{6.8}$$

Here σ_i^k is the Euclidean distance between the objective values of the k-th and the i-th individual. The value for k is equal to the square root of the sample size: $k = \sqrt{N + \overline{N}}$ (20). The values N and $\overline{N}$ denote the number of individuals in the population and archive respectively.

Selection

The selection process produces a new population from the current population and the archive using binary tournament selection (39). Two individuals are picked randomly without replacement from either the population or the archive. The probability of selecting an individual from the archive rather than the population is determined by the 'elitism degree' (ED) (32) (see Equation 6.9).

$$ED = \begin{cases} 1 - \frac{|P_t|}{|\overline{P_t} \cup P_t|} & \text{if } |\overline{P_t}| \geq 2 \\ 0 & \text{otherwise.} \end{cases} \tag{6.9}$$

Here $|P_t|$ is the size of the current population and $|\overline{P_t} \cup P_t|$ is the size of the archive and the current population. Hence, the larger the archive is, the more likely it is that an individual is sampled from the archive. The individual with the lowest fitness value (see Equation 6.5) is declared as the winner and inserted into the new population. If a tie occurs, one individual is chosen with a uniform probability. This procedure is repeated until the new population has reached the size of the old one.

The Variation Operators

The MOEA deploys several variation operators (VOs) as it may produce superior results (29). Different variation operators are also necessary due to the complex structure of the representation scheme. For example, one requires different operators to change the number of rules, number of features, and the parameters of the membership function. Of course, this could be achieved by a single operator that reinitializes the individual. However, this would be too disruptive.

Two types of VOs were implemented. The mutation operator (MO) changes the structure of one individual and the crossover operator (CO) exchanges rules and/or their parts between two individuals. The probability of applying a MO to an individual is determined by the mutation probability value (MP), which is encoded in the individual as a bit string. If a random generator produces a value less than the MP value, a particular MO is chosen with a uniform probability from the possible MOs. To apply several COs, two individuals are selected without replacement from the population. The probability of applying a CO is determined by the average CP (crossover probability) values encoded within each individual. If it is decided that a CO is applied, a particular CO is chosen with a uniform probability from the possible COs. This procedure is repeated until there are less than two individuals within the population (one or none). The resulting population is then passed to the fitness evaluation process (see Figure 6.1). More details about the implemented VOs are provided in (63).

The Self-Adaptation Scheme

An empirical study has shown that significant interactions exist between different parameters of the system, and that each data set requires different parameter values (65). Due to the large number of parameter combinations it is impractical to determine optimal parameters for each new data set. Hence, the implemented MOEA deploys self-adaptive parameters. This makes the MOEA more practical and possibly more effective and efficient, as the parameters are not static during the evolutionary process. The deployment of static parameters can be disadvantageous as different stages of the evolutionary search may require other parameter values (43). The afore-mentioned probabilities MP and CP are encoded in the individual as bit strings. Standard genetic algorithm variation operators (e.g. (7)) are applied to these bit strings before the actual variation operators are applied to the FCRSs. Hence, the MOEA does not only search for good problem solutions but also (implicitly) for optimal parameter values. This is because individuals with optimal parameters and objective values are more likely to survive the selection process.

The Archive

An archive is a crucial component of a MOEA but standard archiving strategies do not guarantee convergence and population diversity (30). Consequently, we deployed a new archiving strategy that was proposed in (30). It ensures convergence and the diversity promotion. The interested reader is referred to (30) for a detailed description.

6.2.3 Results and Discussion

This Section compares the implemented MOEA with two studies that evaluated several inducers using AUC-based measures on data sets from the UCI Machine Learning Repository[2]. The implemented MOEA has two parameters: *population size* and *number of generations*. Both parameters were fixed to a value of 150 and 1000, respectively. The FCRSs could contain a maximum of 100 rules. The maximum number of antecedent nodes depended on the number of features within each data set. As the MOEA generates several trade-off solutions, the FCRS with the largest AUC value on the validation data set is chosen from the archive as the final output. If there are several FCRSs with the same AUC values, the FCRS with the fewest rules and nodes is chosen. Ten-fold stratified cross-validation was used to generate ten validation data. Table 6.2 compares the implemented MOEA with two approaches reported in (66). The first approach, EROL uses an evolutionary algorithm proposed in (66) whereas SVM is a support vector machine approach (e.g. (40)).

The last three columns contain the average AUC values together with the standard deviations, computed according to (54) using ten-fold cross-validation. The results show that the MOEA outperforms both approaches four out of seven times (in terms of the average AUC values).

[2] http://www.ics.uci.edu/~Emlearn/MLRepository.html

Table 6.2. Comparison of the MOEA with two existing inducers. The average AUC values (computed according to (54) using ten-fold cross-validation) are reported, together with the standard deviations

Data set	EROL	SVM	MOEA
Bcw	67.39 ± 5.10	67.19 ± 5.30	99.29 ± 0.90
Crx	81.63 ± 5.60	83.92 ± 4.40	93.48 ± 4.22
German	71.20 ± 3.50	69.03 ± 2.30	75.61 ± 2.60
Promoters	86.26 ± 6.80	97.44 ± 1.60	93.13 ± 6.84
Vehicle	99.45 ± 0.53	99.33 ± 0.72	90.16 ± 2.24
Votes	99.29 ± 0.40	98.86 ± 0.50	99.00 ± 1.10
Waveform	97.07 ± 0.38	96.31 ± 7.80	91.46 ± 0.68

Table 6.3 compares the MOEA with the best results achieved by three inducers reported in (28). The RL approach (34; 41) deploys a general-to-specific beam search for rules, which attempts to reduce the complexity of an exhaustive search by only concentrating on a limited number of promising solutions. The Naive Bayes approach performs classification by estimating the posterior probabilities (e.g. (5)). The *C4.5rules* approach generates rules by post processing decision trees produced by the C4.5 program (18). As both the RL and the C4.5rules method are producing rules, the average number of generated rules is also reported. Please note that all AUC values in Table 6.3 were computed according to (28) in order to allow a comparison between the MOEA and the other approaches. The method proposed in (28) differs only slightly from that in (54).

Table 6.3. Comparison of the MOEA with three other approaches. The average number of rules and average AUC values are reported with their standard deviations

	RL		Naive	C4.5rules		MOEA	
Data set	**Best**	**Rules**	**Bayes**	**Results**	**Rules**	**Results**	**Rules**
Bcw	97.6 ± 1.3	306.5	93.1 ± 5.5	97.6 ± 3.0	8.2	99.40 ± 0.62	13.0
Car	94.3 ± 1.4	107.6	92.3 ± 2.2	98.4 ± 0.7	78.6	95.39 ± 0.38	40.57
Cmc	63.9 ± 4.0	196.6	64.1 ± 5.8	67.0 ± 2.6	39.1	71.14 ± 1.39	27.3
Crx	90.2 ± 4.2	758.5	87.6 ± 4.3	90.9 ± 2.6	12.9	94.00 ± 3.37	10.1
German	71.9 ± 4.9	807.5	77.1 ± 4.5	68.2 ± 9.0	23.4	79.06 ± 2.58	20.8
Glass	74.4 ± 10.0	183.7	74.0 ± 8.7	77.1 ± 5.2	12.2	94.4 ± 1.55	48.0
Image	93.3 ± 1.4	811.4	95.6 ± 0.9	99.1 ± 0.5	28.6	97.8 ± 1.62	25.1
Kr-vs-kp	92.6 ± 1.5	2328.3	95.1 ± 0.8	99.9 ± 0.1	26.3	90.39 ± 4.45	5.9
Mushroom	100.0 ± 0.0	2362.2	99.8 ± 0.1	100.0 ± 0.0	11.5	99.77 ± 0.17	7.5
Nursery	97.1 ± 0.2	606.6	98.0 ± 0.2	99.8 ± 0.0	336.8	97.04 ± 0.3	34.8
Promoters	83.5 ± 16.2	7432.2	97.7 ± 4.0	89.9 ± 11.2	8.0	93.22 ± 6.7	15.9
Sonar	65.8 ± 12.8	10075.7	76.1 ± 13	77.8 ± 12.3	9.1	83.06 ± 6.65	31.4
Splice	87.3 ± 1.6	8406.8	99.2 ± 0.6	97.5 ± 0.6	76.2	91.54 ± 1.51	20.3

Table 6.3 shows that the MOEA outperforms the RL approach ten out of thirteen times, the Naive Bayes approach eight out of thirteen times, and the C4.5rules approach seven out of thirteen times in terms of the average AUC values. The MOEA also outperforms the RL approach thirteen out of thirteen times and the C4.5rules approach nine out of thirteen times in terms of the average number of rules. This emphasizes the fact that the implemented MOEA produces FCRSs that fit the data well, but are also comprehensible because they contain fewer rules.

6.3 Survival Analysis Using Multi-Objective Evolutionary Algorithms

One aim of survival analysis is to model life/failure time distributions to approximate the probability of an event occurring to a system (e.g. patient) depending on the time and the system's features (10). Features could be, for example, demographic information and/or physiological information. An event may be the recurrence of a disease or death. To model lifetime distribution, or conversely the probability of survival, has a number of benefits. It allows clinicians to devise suitable treatment regimes and counsel patients about their prognosis. Hence, it helps patients to plan their lives and provide future care for their dependents. Survival analysis is also widely used in the social and economic sciences, as well as in engineering. Here the systems being observed are, for example, machines or customers. An event might be the failure of a machine or the churning of a customer. Survival analysis is therefore also referred to as failure time and reliability analysis in these domains (11; 12). Another important task of survival analysis is to test whether different (patient) groups exhibit significantly different lifetime distributions. This Section introduces two MOEAs. One MOEA was used to estimate lifetime distributions and the other to determines groups with different lifetime distributions. The automatic determination of groups with different lifetime distributions is tremendously important. It may help to assign treatments and limited resources correctly and thus save life and money. Furthermore, it may generate new insights. Please note that this Section only provides brief summaries of the MOEA (see (22; 50) for more details).

6.3.1 Introduction to Survival Analysis

Table 6.4 contains a data set taken from (51) of 42 leukaemia patients with one dichotomous feature that indicates whether or not the patient received a particular treatment.

The time is measured in weeks, and the maximum time horizon of the study (the follow up time) is 35 weeks. Survival times without plus signs indicate patients who died, whereas survival times with plus sign indicate patients who were censored. In essence, censoring occurs if one knows the status of the patient only for a

Table 6.4. Leukaemia data taken from (51). The numbers correspond to the survival times in weeks after the patient entered the study. The plus sign indicates censoring

Group 1 (Treatment)	**Group 2** (Placebo)
6, 6, 6, 7, 10, 13, 16, 22, 23, 6+, 9+, 10+, 11+, 17+, 19+, 20+, 25+, 32+, 32+, 34+, 35+	1, 1, 2, 2, 3, 4, 4, 5, 5, 8, 8, 8, 8, 11, 11, 12, 12, 15, 17, 22, 23

particular period of time, but not for the complete follow up time (10). It has to be noted that censoring makes survival analysis problems a challenging task and different to classification problems. Figure 6.4 depicts the survival probability curves of the two patient groups that were estimated using the Kaplan-Meier method (see Equation 6.10).

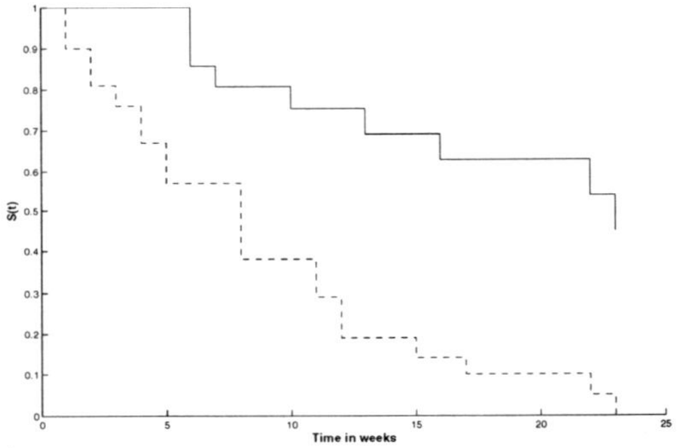

Fig. 6.4. Kaplan-Meier curves for the Leukaemia data (treated patients - solid line, untreated patients dashed line)

$$\hat{S}(t) = \prod_{j=1}^{k}(\frac{n_j - d_j}{n_j}) = \prod_{j=1}^{k}(1 - \frac{d_j}{n_j}) = \hat{S}(t-1)(1 - \frac{d_t}{n_t}). \qquad (6.10)$$

Here d_j corresponds to the number of events (*e.g.* deaths) at time j where events occurred and n_j corresponds to the number of objects that are still observed at time j. The survival probability (6.11) is related to the hazard function (6.12) (11).

$$S(t,x) = e^{-\int_0^t h(u)\,du}. \qquad (6.11)$$

$$h(t,x) = -\frac{d}{dt}\,log(S(t)). \qquad (6.12)$$

Figure 6.4 shows that the treatment group has a better survival than the placebo group. This difference can be quantified using the log-rank statistic. The leukaemia data have a value of 16.79 when grouped according to the dichotomous feature *treatment*). The computation of the log-rank statistic for G groups is summarized in Equation 6.13.

$$L = d^{'} V^{-1} d. \tag{6.13}$$

Vector d corresponds to the sum of the k vectors for the event times: $j = 1, 2, \ldots, k$.[3] The $G-1$ values for each vector are computed according to Equation 6.14.

$$d_j = (O_1 - E_1, O_2 - E_2, \ldots, O_{G-1} - E_{G-1}). \tag{6.14}$$

The difference between observation O_i and expectation E_i for event time j for group i is computed according to Equation 6.15.

$$O_i - E_i = \sum_{j=1}^{k} m_{ij} - (\frac{n_{ij} m_j}{n_j}). \tag{6.15}$$

Here m_{ij} denotes the number of events at event time j in group i, n_{ij} the number of samples at risk at event time j in group i, m_j the number of events at event time j, and n_j the number of sample at risk at event time j. Matrix V corresponds to the sum of k variance matrices for the event times. Each matrix consists of $i = 1, 2, \ldots, G - 1 \times l = 1, 2, \ldots, G - 1$ elements, which are computed according to Equation 6.16.

$$V(i,l) = \begin{cases} \sum_{j+1}^{k} \frac{n_{ij}(n_j - n_{ij}) m_j (n_j - m_j)}{n_j^2 (n_j - 1)} & i = l \\ \sum_{j+1}^{k} = \frac{-n_{ij} n_{lj} m_j (n_j - m_j)}{n_j^2 (n_j - 1)} & i \neq l. \end{cases} \tag{6.16}$$

6.3.2 Artificial Data

Four artificial data sets were created to determine whether the MOEAs can cope with certain data set features, such as non-proportional hazard distributions (crossing survival probability curves), interaction effects and noise. These features can cause problems for classical approaches such as the standard Cox model.

Artificial Data Set 1

This artificial data set was inspired by the well-known XOR classification problem. It has two binary features: $X0$ and $X1$ and a third binary feature that has to be predicted. Two features replaced the third feature in order to simulate a survival

[3] If there are more than two groups, d corresponds to a vector and V to a matrix, where $d^{'}$ is the transpose vector and V^{-1} an inverse matrix.

analysis problem rather than a classification problem. The first feature (I) indicated whether the event occurred to the sample and the second feature ($Time$) determined the observation time of the sample. The maximum observation time was set to a value of ten. The actual data consisted of 50 samples for each feature value combination (see Table 6.5).

Table 6.5. Feature value combinations (two left columns) and inverse lognormal distribution parameters (two right columns)

x_0	x_1	μ	σ
0	0	1.1	0.15
0	1	1.8	1.1
1	0	1.8	1.1
1	1	1.1	0.15

Two values were generated to obtain the observation time for a sample. The first value (α) was sampled from the interval $[0 \ldots 10]$ with a uniform probability. The second value (β) was sampled from a inverse lognormal distribution (13), which is characterized by the parameters μ and σ. The parameter values for a sample (feature value combination) are summarized in Table 6.5. To simulate censoring the actual observation time and the indicator value were determined according to Equation 6.17.

$$(I, Time) = \left| \begin{array}{ll} (1, \beta) & for\ \alpha \geq \beta \\ (0, \alpha) & \text{otherwise.} \end{array} \right. \tag{6.17}$$

Artificial Data Set 2

This artificial data set was generated in the same manner as the first artificial data set. However, it contains an additional 'noise' feature, which was generated by sampling from the interval $[0 \ldots 1]$ with a uniform probability. As the data set contained more features, 100 instead of 50 samples were produced for each feature combination.

Artificial Data Set 3

This artificial data set was generated in a similar manner as the first artificial data set. It contains three features that can have the value one or zero resulting in eight possible combinations as shown in Table 6.6. We produced 150 samples for each feature value combination.

Table 6.6. Feature value combinations (two left columns) and inverse lognormal distribution parameters (two right columns)

x0	x1	x2	μ	σ
0	0	0	1.10	0.15
0	0	1	3.50	1.70
0	1	0	1.80	1.10
0	1	1	3.50	1.70
1	0	0	1.80	1.10
1	0	1	3.50	1.70
1	1	0	1.10	0.15
1	1	1	3.50	1.70

Artificial Data Set 4

This artificial data set was originally proposed in (53) and uses a mixture of a Weibull and Gamma distribution. The distribution parameters depend on the values of the two features x_1 and x_2 (6.18).

$$f(t|x) = x_2\, W(x_1^2, sin(x_1)^3 + 2) + (1 - x_2) \cdot G((x_1 + 1)^2, exp(cos(x_1) + 2)). \tag{6.18}$$

The values of x_1 and x_2 were first sampled from a bivariate Gaussian with the mean vector $(0, 0)$ and the covariance $[1.0 \;\; -0.5; 1.0\; 0.5]$. The variable x_2 was then transformed to a binary indicator (values greater than or equal to zero were set to one, values less than zero were set to zero). Five hundred training and test data samples were produced. Each sample was censored with a probability of 0.27 (53).

6.3.3 Medical Data Set

The medical data set contains feature values of 1820 uveal melanoma patients (15) (see Table 6.7). Uveal melanomas have an occurrence rate of six per million per year (48) and arise from melanocytes in the uvea. Patients with uveal melanoma usually have symptoms such as blurred vision, flashing lights and visual field loss. Without treatment many eyes become blind, painful and cosmetically unsightly. Approximately 50% of all patients with uveal melanoma ultimately die of this disease, nearly always as a result of haematogenous spread of tumor to the liver.

6.3.4 Survival Probability Estimation Using a Multi-Objective Evolutionary Algorithm

The implemented MOEA exploits the fact that the hazard corresponds to a conditional probability in the discrete time domain as shown in Equation 6.19 (14; 49).

Table 6.7. Features of the uveal melanoma data set

Name	Type	Description
Antora	Dichotomous	Indicates whether the tumor is at the front or the back of the eye (anterior choroid or posterior choroid).
Age	Continuous	Age of the patient when (s)he entered the study.
Ludb	Continuous	Tumor dimension as measured by ultrasonography.
Gender	Dichotomous	n/a

$$\hat{h}(t_j, x) = Pr(T = t_j | T \geq t_j, x). \tag{6.19}$$

In the discrete case, the survival probabilities can be computed according to Equation 6.20 for each time interval t_j.

$$\hat{S}(t_j, x) = \prod_{k=1}^{j} (1 - \hat{h}(t_j, x)_k). \tag{6.20}$$

It follows that the original survival analysis problem can be cast as a classification problem that that can be estimated using conditional probability. However, this requires pre-processing of the data due to the problem of censoring. The resulting data consist of variables for the features, time and the event/indicator (see (50) for more details). The event has to be predicted. This can be achieved by estimating the conditional probability $P(Event|x)$ (x represents the feature and time variables) using, for example, logistic regression models or artificial neural networks (ANNs) (49; 60). An advantage of ANNs is that they can directly produce smooth estimates for the discrete hazard without estimating the baseline hazard. In addition, ANNs can cope with non-proportional hazards and interaction effects. However, this can also lead to overfitting. This work fitted radial basis function networks (RBFNs) to eliminate problems associated with ANNs. For example, RBFNs have simpler structures which leads to simpler search spaces and shorter extraction times (9). In addition, the parameters of a RBFN can be easier interpreted. This addresses one crucial problem of ANNs, namely their difficult interpretability (61). The implemented MOEA has the same structure as described in Section 6.2.

The Representation Scheme

Equation 6.21 summarizes the the structure of a RBFN.

$$\xi_k = \sum_{j=1}^{M} w_{kj} z_j(x) + b_k. \tag{6.21}$$

Here w_{kj} are coefficients (weights), z_{kj} is the output of the j-th basis functions, and b is a bias. The index k denotes the predicted class. This index is only necessary if one intends to predict more than two classes. The present study used a Gaussian basis function. Note that other basis functions (kernels) could also be used. The implemented representation scheme consists of basis functions (Equation 6.21), which are represented as trees (Figure 6.3). Non-terminal nodes (circles in Figure 6.3) correspond to the product operator. The terminal nodes (squares in Figure 6.3) correspond to Gaussian basis functions. A tree can contain at most one basis function for each feature. Note that neither the number of basis functions nor the number of features is preset. They can be between a minimum and maximum value enabling the MOEA to adapt the complexity of the RBFN. This also means that the MOEA can perform an implicit feature selection, because it is not forced to use all features within the trees. Each tree also has k associated weight vectors, which contain values between zero and one (Equation 6.21). The k-th bias is a real number.

Fitness Evaluation

The MOEA uses the fitness assignment described in Section 6.2.2 and minimizes three objectives: fit of the RBFN to the data, number of basis functions and number of different features. The fit of the RBFN to the data is measured using the cross-entropy, which is computed according to Equation 6.22 (9).

$$E = -\sum_{i=1}^{N}\sum_{j=1}^{L}\{d_{ij}log[h(t_j,x)] + (1-d_{ij})log[1-h(t_j,x)]\}. \tag{6.22}$$

Here d_{ij} is the event/indicator (for the i-th vector for the j-th time interval. The hazard $h(t_j,x)$ is the output of the model.

Variation Operators

Several variation operators were implemented (see (50) for further details).

Results

The MOEA is evaluated on the leukaemia data set (51) and four artificial data sets (see Section 6.3.2 and the medical data set (Section 6.3.3). Data sets were randomly split into a training data set and test data set for each experiment. Training data sets contained two-thirds and the test data set one-third of the original data set. Before the MOEA was run a holdout data set was randomly selected from the training data set. It contained one third of the training data set. The MOEA was then run and the generated individuals in the archive and the population were re-evaluated on the holdout data set. As each run of the MOEA produces several trade-off solutions, the RBFN with the best fit to the holdout data set (see Equation 6.22) was chosen as the final output of the algorithm. It should be noted, that if there were several RBFNs with the same fit to the validation data, the RBFN with the smallest number of basis functions and features was chosen. To choose the model with the best fit to the holdout data set rather than the training data set makes it more likely that the model generalizes on unseen data. This method is also referred to as holdout method (9). The following parameters were used during the experiments: Population Size 100, Number of Generations 500, Crossover Probability - 0.7, Mutation Probability - 0.3. Figure 6.5 shows the Kaplan-Meier estimates and the model estimates for the Leukaemia data set.

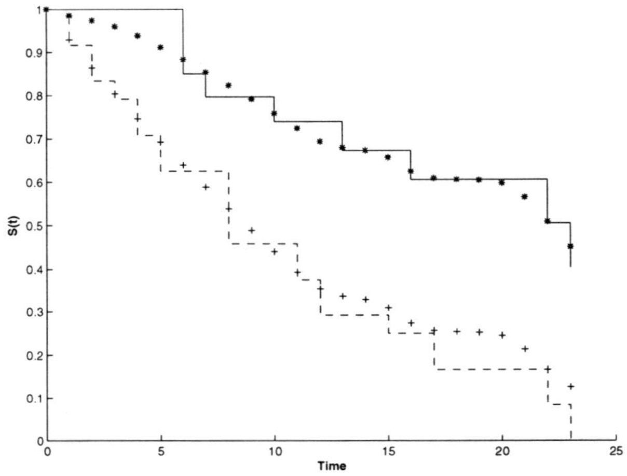

Fig. 6.5. Kaplan-Meier estimates for the Leukaemia data set line - treated patients, dashed line untreated patients) together with the estimates of the generated model (stars - treated patients, crosses - untreated patients)

Figure 6.6 shows the responses of the evolved RBFN for the first artificial data set. Each feature value combination is shown together with the 'true' survival functions.

The r-square measure was used to measure the correlation between the expected values and the predicted values. The model achieved an r-square value of 0.9154. This shows that the model predicts the survival functions correctly and that the MOEA can be applied to survival data with non-proportional hazard distributions. The RBF for the second artificial data set achieved an r-square measure of 0.9219 and excluded the 'noise' variable. This shows that the implemented MOEA can also be applied to noisy hazard distributions. Figure 6.7 depicts the estimations of the model for the third artificial data set.

The evolved model achieves an r-square value of 0.9491. Figure 6.8 shows the Kaplan-Meier estimates and the average survival function estimates of the evolved

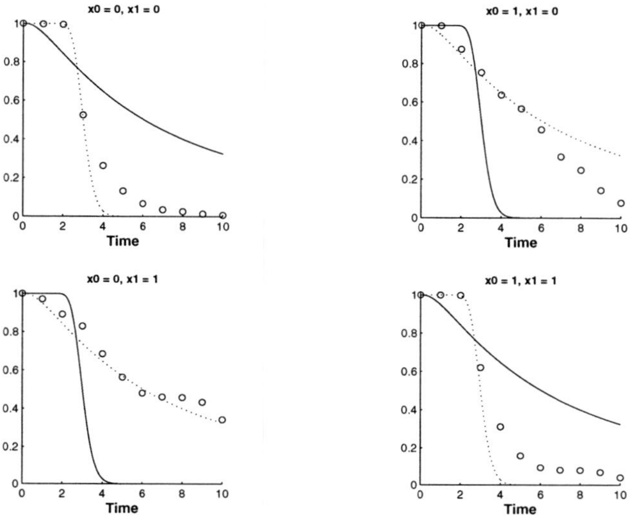

Fig. 6.6. Response values of the model for each feature value combination. The correct survival function is shown as dotted line

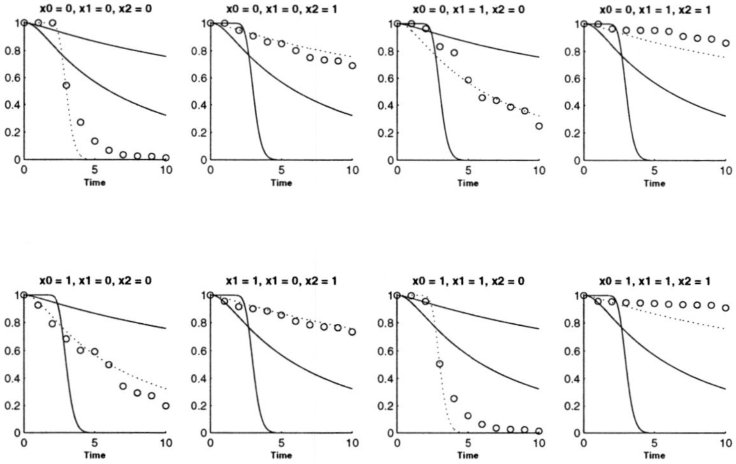

Fig. 6.7. Estimations of for each feature value combination (circles). Dotted lines depict the 'true' survival functions for the particular feature value combination

RBFN for each group. The estimates of the evolved RBFN at $t = 8^4$ had to be divided into five groups of equal size and were ordered.

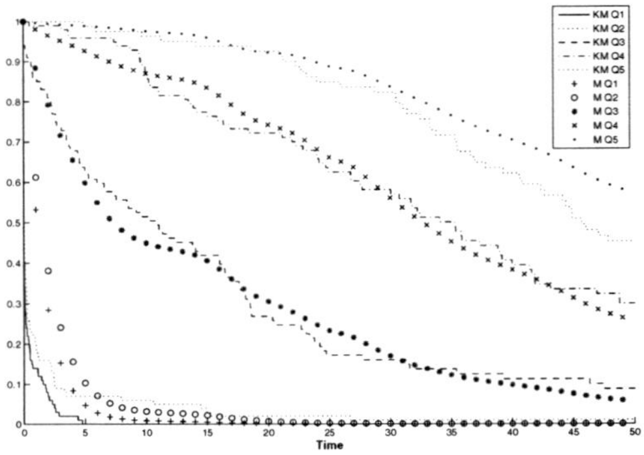

Fig. 6.8. Kaplan-Meier estimates (KM Q1 - KM Q5) and average survival function estimates of the evolved model (M Q1 - M Q5) for all five groups

All models captured the distribution of the data because the estimates and the "underlying truth" agree (53).

6.3.5 Discovery of Groups with Different Lifetime Distributions

The implemented MOEA fits oblique decision trees that divide the feature space into mutually exclusive subsets whilst maximizing the log-rank statistic. The structure of the MOEA is the same as described in Section 6.2.

6.3.6 Representation Scheme

A tree shaped representation scheme is used to represent oblique decision trees. Each non-terminal node (circles in Figure 6.3) corresponds to a linear combination of the features. It returns the value of its left child, if the value of the linear combination is greater or equal zero; otherwise it returns the value of the right child. Terminal nodes (squares in Figure 6.3) correspond to integer values that indicate the group membership. The tree is initialized randomly. A predefined lower and upper bound restrict the number of nodes. The coefficients of the linear combination are sampled from a predefined interval with a uniform probability. The representation scheme also contains a binary vector that determines whether or not a feature is used within the linear combinations of the non-terminal nodes enabling the algorithm to perform implicit feature selection.

[4] The median survival time is 8.4 (53).

6.3.7 Fitness Evaluation

The MOEA uses the same fitness assignment described in Section 6.2.2 and minimizes four objectives of the decision trees: log-rank statistic (Equation 6.13), number of features, number of nodes, number of distinct groups/terminal node.

Variation Operators

The MOEA uses several variation operators that work on different parts of the representation scheme to achieve an appropriate exploitation and exploration of the search space (see (22) for more details).

Results

The implemented MOEA is evaluated on the first three artificial data sets (Section 6.3.2 and one medical data set (Section 6.3.3). The model with the maximum log-rank statistic (Equation 6.13) was chosen as the final output of the algorithm. If there were several models with the same log-rank statistic, the model with the smallest number of nodes and group assignments (objective 4) was chosen. The following parameters were used during the experiments: population size 100, number of generations 1000, crossover probability - 0.7, mutation probability - 0.4, minimum number of tree nodes 3, maximum number of tree nodes - 21. Kaplan-Meier curves were produce according to the grouping of the model. Figure 6.9 depicts the Kaplan-Meier curves of the model for the first artificial data set.

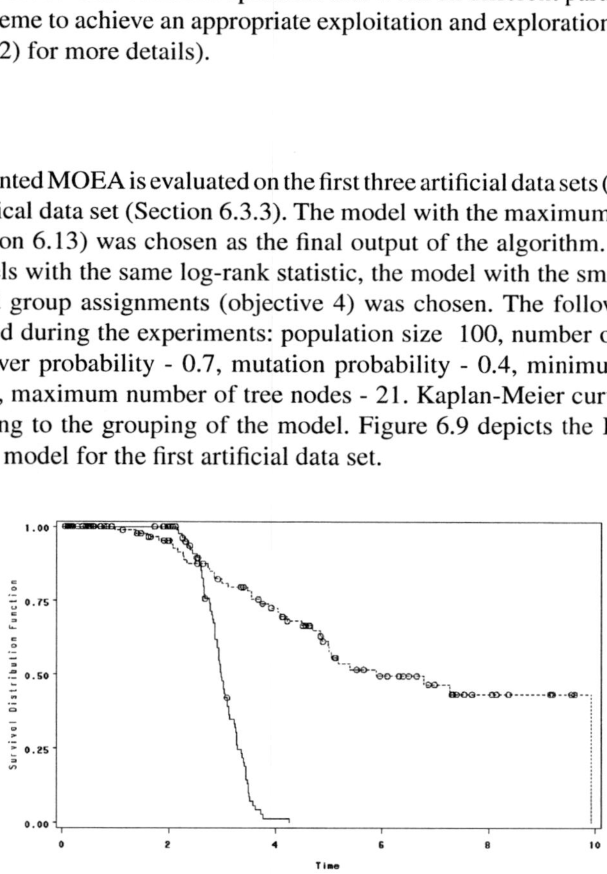

Fig. 6.9. Kaplan-Meier curves according to the grouping of the model

The model achieved a log-rank statistic of 80.76 and recovered the original grouping of the data accurately (see Figure 6.6 for the 'true survival functions). The generated model for the second artificial data set achieved a log-rank statistic of 202.51. It also recovered the underlying grouping of the data correctly and the additional 'noise' feature was dropped. Figure 6.10 shows the Kaplan-Meier curves of the model for the third artificial data set (see Figure 6.7 for the 'true survival functions).

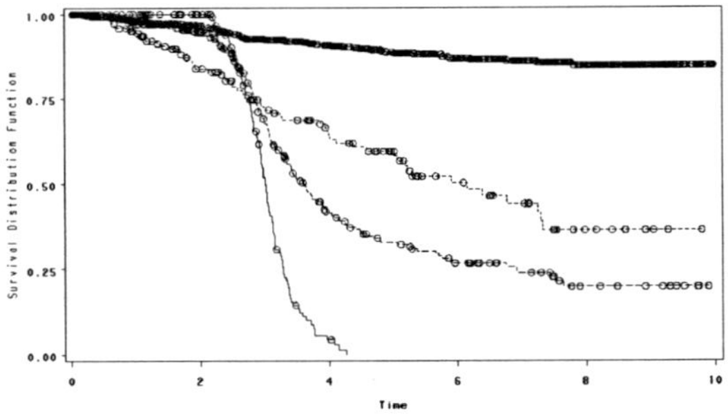

Fig. 6.10. Kaplan-Meier curves according to the grouping of the generated model

The model achieved a log-rank statistic of 425.01. However, 12.5% of the samples were grouped incorrectly. Samples with the feature combination 110 were wrongly assigned to the group with the parameters $\mu = 1.80$ and $\sigma = 1.10$. Furthermore, samples with the feature combination 010 and 100 were assigned to two different groups, although they belong to the same (parameter group) group. The generated model for the medical data set achieved a log-rank statistic of 204.31. It only uses two of the original four features: *Antora* and *Ludb*. This agrees with the medical consultant who provided the data and deemed the features *Antora* and *Ludb* as more important. Figure 6.11 depicts the Kaplan-Meier curves according to the grouping of the generated model.

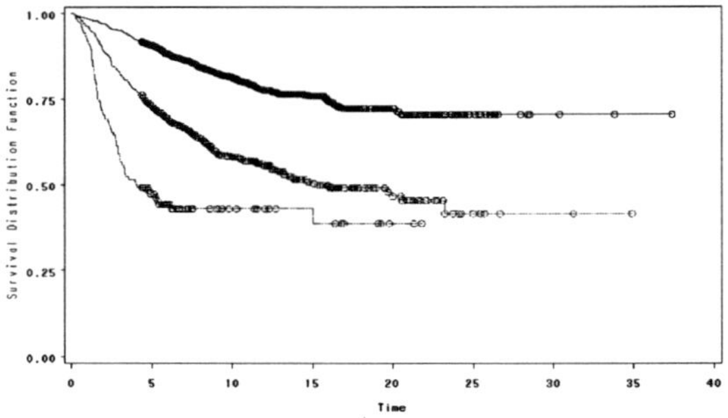

Fig. 6.11. Kaplan-Meier curves according to the grouping of the model

The decision surface of the model is depicted in Figure 6.12. The best Kaplan-Meier curve corresponds to samples that were assigned to group '1' (1060 samples). The second best Kaplan-Meier curve corresponds to samples that were assigned to group '2' (636 samples). The worst Kaplan-Meier curve corresponds to samples that were assigned to group '3' (124 samples).

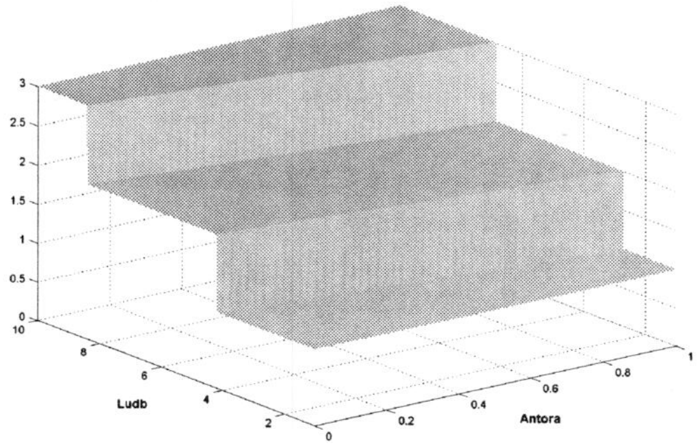

Fig. 6.12. Decision surface of the generated model

The medical consultant agreed with this decision surface. However, whether or not the model is better than other models, such as tumor nodes metastasis staging (TNM) (2), remains to be investigated.

6.4 Summary

This chapter discussed and evaluated three MOEAs. The first MOEA was used for the induction of fuzzy classification rule system. The other two MOEAs were used for two survival analysis problems, which are data mining tasks that have not previously been tackled with MOEAs. We focused on the induction of comprehensible models whilst overcoming several shortcomings of existing approaches. Furthermore, we evaluated the implementations on several benchmark and artificial data. This chapter only provided a brief overview of the implemented MOEAs and related subjects. The interested reader is referred to (22; 50; 52; 63) for further details.

References

[1] Freitas A A, Rozenberg G (2002) Data mining and knowledge discovery with evolutionary algorithms. Springer Verlag, Berlin, Heidelberg, New York

[2] Sobin L H, Wittekind C (2002) Classification of malignant tumours. Wiley-Liss

[3] Russell S J, Norvig P (1994) Artificial intelligence: A modern approach. Prentice Hall
[4] Holland J H (1975) Adaptation in natural and artificial systems: an introductory analysis with applications to biology, control, and artificial intelligence. University of Michigan Press, Ann Arbor
[5] Michie D, Spiegelhalter D J, Taylor C C (1994) Machine learning, neural and statistical classification. Ellis Horwood
[6] Krause P, Clark D (1993) Representing uncertain knowledge: An artificial intelligence approach. Kluwer Academic Publishers
[7] Michalewicz Z, Fogel D B (2005) How to solve it: modern heuristics. 2nd Edition, Springer, Berlin
[8] Gordon A D (1981) Classification. Chapman and Hall
[9] Bishop C M (1995) Neural networks for pattern recognition. Oxford University Press, Oxford
[10] Kleinbaum D G (1996) Survival analysis: A self-learning text. Springer
[11] Afifi A, Clark V A, May S (2003) Computer-aided multivariate analysis. Chapman and Hall
[12] Kalbfleisch J D, Prentice R L (1980) The statistical analysis of failure time data. Wiley
[13] Evans M, Hastings N, Peacock B (1993) Statistical distributions. John Wiley and Sons
[14] Lawless J F (2002) Statistical models and methods for lifetime data. Wiley, New York
[15] Damato B E (2000) Ocular tumours : diagnosis and treatment. Butterworth Heinemann
[16] Klir G J, Clair U S, Yuan B (1997) Fuzzy set theory: foundations and applications. Prentice Hall, Upper Saddle River, NJ
[17] Deb K (2001) Multi-objective optimization using evolutionary algorithms. Wiley Chichester
[18] Quinlan J R (1994) C4.5 : programs for machine learning. Morgan Kaufmann
[19] Duda R O, Hart P E, Stork H G (2000) Pattern classification. Wiley-Interscience, New York
[20] Silverman B W (1999) Density estimation for statistics and data analysis. Chapman and Hall
[21] Fayyad U M, Piatetsky-Shapiro G, Smyth P (1996) From data mining to knowledge discovery in databases. Advances in Knowledge Discovery and Data Mining, AAAI Press/The MIT Press 1–36
[22] Setzkorn C, Taktak A F, Damato B (2006) Evolving oblique decision trees for survival analysis. Poster Proceedings of the 6th Industrial Conference on Data Mining, Springer 144–158
[23] Freitas A A (2000) Evolutionary algorithms. Handbook of Data Mining and Knowledge Discovery
[24] Provost F, Fawcett T, Kohavi R (1998) The case against accuracy estimation for comparing induction algorithms. Proc. 15th International Conf. on Machine Learning, Morgan Kaufmann, San Francisco, CA 445–453

[25] Fonseca C M, Fleming P J (1995) Multiobjective genetic algorithms made easy: selection, sharing, and mating restriction. Proceedings of the First International Conference on Genetic Algorithms in Engineering Systems: Innovations and Applications, Sheffield, UK 42–52

[26] Holland J H, Reitman J S (2002) Cognitive systems based on adaptive algorithms. In: D. A. Waterman and F. Hayes-Roth (eds) Pattern Directed Inference Systems, Academic Press 313–329

[27] Holland J H (1986) Escaping brittleness: The possibility of general-purpose learning algorithms applied to rule-based systems. In: R. S. Michalski and J. G. Carbonell and T. M. Mitchell (eds) Machine Learning: An Artificial Intelligence Approach, Volume II, Morgan Kaufmann 593–623

[28] Fawcett T (2001) Using rule sets to maximize ROC performance. Proceedings of the IEEE International Conference on Data Mining, IEEE Computer Society 131–138

[29] Spears W M (1995) Adapting crossover in evolutionary algorithms. In: J. R. McDonnell and R. G. Reynolds and D. B. Fogel (eds) Proc. of the Fourth Annual Conference on Evolutionary Programming, MIT Press, Cambridge, MA 367–384

[30] Laumanns M, Thiele L, Zitzler E, Deb K (2002) Archiving with guaranteed convergence and diversity in multi-objective optimization. In: Langdon W B, Cantú-Paz E, Mathias K, Roy R, Davis D, Poli R, Balakrishnan K, Honavar V, Rudolph G, Wegener J, Bull L Potter M A, Schultz A C, Miller J F, Burke E, Jonoska N (eds) GECCO 2002: Proceedings of the Genetic and Evolutionary Computation Conference, Morgan Kaufmann Publishers, San Francisco, CA 94104, USA 439–447

[31] Setzkorn C, Paton R C (2004) JavaSpaces-an affordable technology for the simple implementation of reusable parallel evolutionary algorithms. In: López J A, Benfenati E, Dubitzky W (eds) Knowledge Exploration in Life Science Informatics - KELSI 2004 (LNAI 3303), Springer-Verlag, New York 151–161

[32] Laumanns M, Zitzler E, Thiele L (2000) A unified model for multi-objective evolutionary algorithms with elitism. Proceedings of the 2000 Congress on Evolutionary Computation (CEC 2000). IEEE Press, Piscataway, New Jersey 46–53

[33] Hoffmann F, Pfister G (1995) A new learning method for the design of hierarchical fuzzy controllers using messy genetic algorithms. Proceedings of the Sixth International Fuzzy Systems Association World Congress (IFSA'95). Sao Paulo, Brazil 249–252

[34] Clearwater S, Provost F (1990) RL4: A tool for knowledge-based induction. Proceedings of the Second International IEEE Conference on Tools for Artificial Intelligence 24–30

[35] Duch W, Jankowski N, Grabczewski K, Adamczak R (2000) Optimization and interpretation of rule-based classifiers. Intelligent Information Systems, Advances in Soft Computing 1–14

[36] Nürnberger A, Klose A, Kruse R (2000) Analyzing borders between partially contradicting fuzzy classification rules. Proceedings of 19th International

Conference of the North American Fuzzy Information Processing Society (NAFIPS 2000) 59–63

[37] Cramer N L (1985) A representation for the adaptive generation of simple sequential programs. Proceedings of the International Conference on Genetic Algorithms and Their Applications 183–187

[38] Koza J R (1998) Genetic programming. In: Williams James G, Kent A (eds) Encyclopedia of Computer Science and Technology, Marcel-Dekker 29–43

[39] Zitzler E, Laumanns M, Thiele L (2001) SPEA2: Improving the strength pareto evolutionary algorithm. EUROGEN 2001 - Evolutionary Methods for Design, Optimisation and Control with Applications to Industrial Problems 19–26

[40] Burges C J C (1998) A tutorial on support vector machines for pattern recognition. Data Mining and Knowledge Discovery 2(2):121–167

[41] Provost F J, Aronis J M (1996) Scaling up inductive learning with massive parallelism. Machine Learning 23(1):33–46

[42] Dhar V, Chou D, Provost F J (2000) Discovering interesting patterns for investment decision making with GLOWER - A genetic learner overlaid with entropy reduction. Data Mining and Knowledge Discovery 4(4):251–280

[43] Eiben A E, Hinterding R, Michalewicz Z (1999) Parameter control in evolutionary algorithms. IEEE Trans. on Evolutionary Computation 3(2):124–141

[44] Grohman W M, Dhawan A P (2001) Fuzzy convex set-based pattern classification for analysis of mammographic microcalcifications. Pattern Recognition 34:1469–1482

[45] Nauck D, Kruse R (1999) Obtaining interpretable fuzzy classification rules from medical data. Artificial Intelligence in Medicine 16:149–169

[46] Russo M (1997) FuGeNeSys - a fuzzy genetic neural system for fuzzy modeling. IEEE Transactions On Fuzzy Systems 6(3):373–388

[47] Jain A, Duin P, Mao J (2000) Statistical pattern recognition: A review. IEEE Transactions on Pattern Analysis and Machine Intelligence, 22(1):4–37

[48] Damato B E (2005) Current management of uveal melanoma. European Journal of Cancer Supplements 3(3):433–435

[49] Singer J D, Willett J B (1993) It's about time: Using discrete-time survival analysis to study duration and the timing of events. Journal of Educational Statistics 18(2):155–195

[50] Setzkorn C, Taktak A F, Damato B (2006) On the use of multi-objective evolutionary algorithms for survival analysis. BioSystems (in press).

[51] Freireich E J et al.(1963) The Effect of 6-mercaptopurine on the duration of steroid-induced remissions in acute leukemia. Blood 21:699–716

[52] Setzkorn C, Paton R C (2005) On the use of multi-objective evolutionary algorithms for the induction of fuzzy classification rule systems. BioSystems 81(2):101–112

[53] Eleuteri A, Tagliaferri R, Milano L, De Placido S, De Laurentiis M (2003) A novel neural network-based survival analysis model. Neural Networks 16(5-6):855–864

[54] Hand D J, Till R J (2001) A simple generalization of the area under the ROC curve for multiple class classification problems. Machine Learning 45:171–186

[55] Hanley J, McNeil B J (1982) The meaning and use of the area under a receiver operating characteristic ROC curve. Radiology 143:29–36
[56] Bradley A (1997) The use of the area under the ROC curve in the evaluation of machine learning algorithms. Pattern Recognition 30(7):1145–1159
[57] Cordón O, del Jesus M J, Herrera F, Lozano M (1999) A proposal on reasoning methods in fuzzy rule-based classification systems. International Journal of Approximate Reasoning 20:21–45
[58] Steimann F (1997) Fuzzy set theory in medicine. Artificial Intelligence in Medicine 11(1):1–7
[59] Zadeh L A (1965) Fuzzy sets. Information and Control 8(3): 338–353
[60] Biganzoli E, Boracchi P, Mariani L, Marubini E (1998) Feed forward neural networks for the analysis of censored survival data: a partial logistic regression approach. Statistics in Medicine 17(10):1169–1186
[61] Clark T G, Bradburn M J, Love S B, Altman D G (2003) Survival analysis part IV: Further concepts and methods in survival analysis. British Journal of Cancer 89:781–786
[62] De Jong K A (1975) An analysis of the behaviour of a class of genetic adaptive systems. PhD thesis, University of Michigan
[63] Setzkorn C (2005) On the use of multi-objective evolutionary algorithms for classification rule induction. PhD Thesis, University of Liverpool, Department of Computer Science Liverpool, United Kingdom
[64] Smith S F (1980) A learning system based on genetic algorithm. PhD Thesis, University of Pittburgh
[65] Setzkorn C, Paton R C (2003) MERBIS - A multi-objective evolutionary rule base induction system. ULCS-03-016, University of Liverpool
[66] Sebag M, Azé J, Lucas N (2004) ROC-based evolutionary learning: application to medical data mining. Artificial Evolution, 384–396

7

Clustering Based on Genetic Algorithms

M.N. Murty, Babaria Rashmin, Chiranjib Bhattacharyya

Department of Computer Science and Automation, Indian Institute of Science, Bangalore 560012, India
{mnm, rashmin, chiru}@csa.iisc.ernet.in

Summary. Clustering is an important abstraction process and it plays a vital role in both pattern recognition and data mining. Partitional algorithms are frequently used for clustering large data sets. K-means algorithm is the most popular partitional clustering algorithm; its fuzzy, rough, probabilistic and neural network are also popular. However, a major problem with the K-means algorithm and its variants is that they may not reach the globally optimal solution of the associated clustering problem. Several stochastic search techniques have been suggested in the past to address this problem. Genetic algorithms (GAs) are attractive to solve the partitional clustering problem. However, conventional GA based solutions may not scale well. A recent proposal in the literature is to use a Quad-tree based algorithm for scaling up the clustering algorithm. Unfortunately this solution does not scale up to handle large dimensional data sets. In this chapter, we explain the GA based clustering approaches and propose an efficient scheme for clustering high-dimensional large-scale data sets using GAs based on the well-known CF-Tree data structure. We also discuss the notion of multi-objective clustering.

7.1 Introduction

Clustering generates a partition consisting of cohesive groups or clusters from a given collection of patterns (12). Representations or descriptions of the clusters formed are used in decision making; classification is one of the popular decision making paradigms. Abstractions in the form of cluster representatives/descriptions are useful in efficient classification (13).

The important steps in the clustering process as depicted in Figure 7.1 are

1. pattern representation: patterns are typically represented as vectors or as sentences or strings of characters in a formal language; here we consider only vectorial representations,
2. definition of an appropriate similarity or dissimilarity function,
3. selecting a clustering algorithm and using it to generate a partition and/or description of the clusters, and

M.N. Murty et al.: *Clustering Based on Genetic Algorithms*, Studies in Computational Intelligence (SCI) **98**, 137–159 (2008)
www.springerlink.com

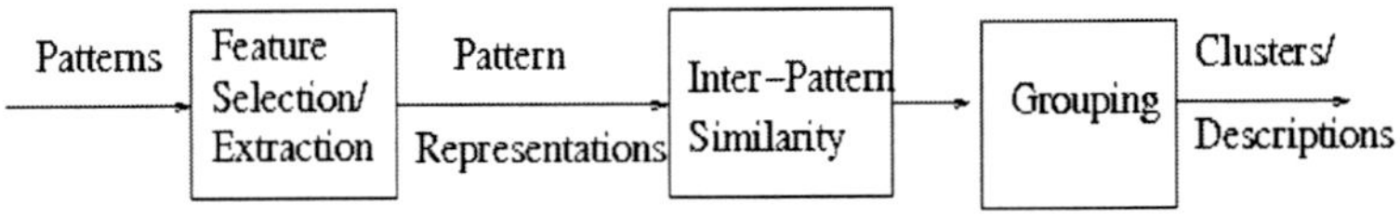

Fig. 7.1. Stages in clustering

4. using these abstractions in decision making.

Clustering has found applications in a large number of areas. These include biometrics, document analysis and recognition, information retrieval, bio-informatics, remote sensed data analysis, bio-medical data analysis, target recognition, and data mining (1). The process of clustering is carried out so that patterns in the same cluster are *similar* in some sense and patterns in different clusters are *dissimilar* in a corresponding sense. This may be illustrated with the help of a two-dimensional data set shown in Figure 7.2. Here, each pattern is represented as a point in the two-dimensional space and there are three clusters. Further, the Euclidean distance between any two points belonging to the same cluster is smaller than that between any two points belonging to different clusters. This notion characterizes similarity; *intra-cluster distance (similarity) is small (high)* and *inter-cluster distance (similarity) is large (low)*. There are several other ways of characterizing similarity.

Given a set of patterns $\mathcal{X} = \{x_1, x_2, \cdots, x_n\}$, the ith cluster $X_i \subset \mathcal{X}$, $i = 1, \cdots, C$, then $\cup_{i=1}^{C} X_i = \mathcal{X}$, and no $X_i = \phi$. If in addition, $X_i \cap X_j = \phi$, for all i and j, $i \neq j$, then we have a hard partition. The number of hard partitions itself could be prohibitively large. For example, there are $2^{n-1} - 1$ 2-partitions, that is partitions having two blocks/clusters, of n patterns. The number of partitions of n patterns into m blocks(clusters) is

$$\frac{1}{m!} \sum_{i=1}^{m} (-1)^{m-i} \binom{m}{i} (i)^n.$$

Consider a very small size problem of grouping 15 patterns into 3 clusters. The possible number of partitions is 2,375,101. So, exhaustive enumeration of all possible partitions to decide the *best* partition, in some sense, is not practical while dealing with large collections of patterns. As a consequence, partitioning is done based on domain knowledge or user provided information in the form of scheme for representation of patterns and clusters, the similarity measure used to characterize inter-pattern similarity, number of clusters, and other parameter values. Further, because of the well-known *theorem of the ugly duckling*, clustering is not possible with out using extra-logical information (2). So, different clustering algorithms use knowledge in different forms to generate the *intended* partition of the data.

Evolutionary algorithms are randomized search techniques based on principles of evolution and natural genetics (3). Some of the well-known evolutionary algorithms

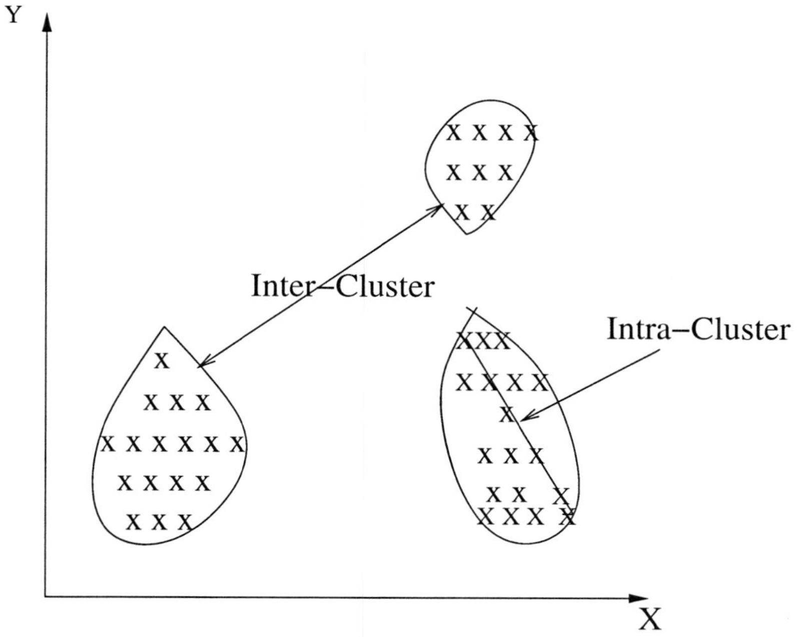

Fig. 7.2. Clustering

(14) are genetic algorithms (GAs), evolutionary algorithms, and evolution strategies. Even though all these paradigms are used in pattern clustering, GAs are used more frequently and are popular in this area. So, we focus on application of GAs to clustering. Here, a clustering problem is viewed as an optimization problem and an evolutionary algorithm is used to search for a globally optimal solution. Conventional clustering algorithms do not guarantee convergence to the globally optimal solution. However, it is possible to show that the evolutionary algorithms converge to a globally optimal solution in a probabilistic sense (5).

The earliest work on this topic is reportedly carried out by Raghavan and Birchard (15). Subsequently, Jones and Beltramo came out with a formulation to solve both hierarchical and nonhierarchical clustering problems using GAs (16). Bhuyan *et al* have shown (17) that GA based clustering can give better results than conventional clustering algorithms. However, based on an empirical study, Mishra and Raghavan have concluded (18) that the performance of GA is not impressive and is inferior to that of other well-known competing search methods like the tabu search. Fogel and Simpson used the evolutionary programming framework (19) to improve the performance of a fuzzy clustering algorithm. Babu and Murty used evolution strategies (20) for both fuzzy and hard centroid based clustering. An good coverage on evolutionary algorithms for clustering is available in the book by Freitas (21). However, most of the GA based clustering algorithms do not scale-up well. They are used in clustering a set of small number of low dimensional patterns. A genetic algorithm using hyper-quadtrees for low-dimensional clustering of large data sets is proposed in (4). In this chapter, we summarize various contributions in literature and propose a novel scheme for GA based clustering using BIRCH (22); the proposed scheme can be used for clustering large high-dimensional data sets.

The chapter is organized as follows. In the next section we describe the background material in the form of clustering and evolutionary algorithms. In Section 3 we describe the existing schemes for GA based clustering. The proposed efficient algorithm for GA based clustering is introduced in Section 4. A discussion on multi-objective clustering is provided in Section 5. Section 6 draws the chapter to conclusion.

7.2 Clustering and Evolutionary Algorithms

There is a wide variety of clustering algorithms. Different approaches can be described with the help of the hierarchy shown in Figure 7.3. Hard clustering algorithms are either hierarchical, where a nested sequence of partitions is generated, or partitional where a partition of the given data set is generated. In this chapter, we focus on the application GAs for clustering which is a soft clustering paradigm. Other soft clustering algorithms (2) are based on fuzzy sets, rough sets, or artificial neural nets (ANNs). Soft clustering generates a set of overlapping clusters.

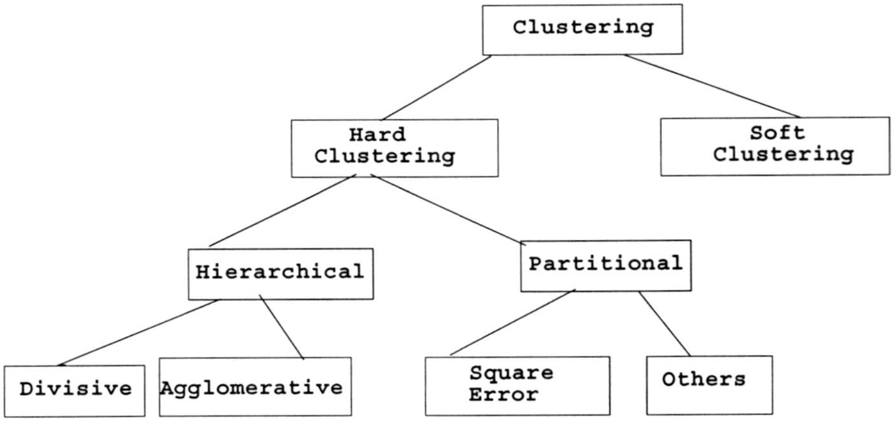

Fig. 7.3. Taxonomy of clustering approaches

7.2.1 Hierarchical Clustering

Hierarchical algorithms produce a nested sequence of partitions of the data which can be depicted by using a tree structure that is popularly called as *dendrogram.*

Hierarchical algorithms are either *divisive* or *agglomerative*. In the former, starting with a single cluster having all the patterns, at each successive step a cluster is split; this process continues till we end up with each pattern in a cluster or a collection of singleton clusters. Conventional hierarchical clustering algorithms are computationally expensive. The agglomerative algorithms require computation and storage of a similarity/dissimilarity matrix of values that has $O(n^2)$ time and space requirement. Initially, they were used in applications where hundreds of patterns were to be clustered. However, when the data sets are larger in size, these algorithms are not feasible because of the non-linear time and space demands. Further, it may not be easy to visualize a dendrogram corresponding to 1000 or more patterns. Similarly, divisive algorithms require exponential time in the number of patterns or the number of features. So, they too do not scale up well in the context of large-scale problems involving millions of patterns/features. However, Balanced Iterative Reducing and Clustering using Hierarchies(BIRCH) (22) provides an efficient hierarchical clustering framework based on a tree structure.

BIRCH employs a data structure called *Cluster Feature Tree* ($CF - tree$ for short). Each CF entry has three values: the number of points in the cluster, the linear sum of the points in the cluster, and the square sum of the data points. In the case of binary data sets, it needs to store only two values in a node as the square sum and linear sum are the same. The CF vector of a cluster C is given by
$CF = (\mid C \mid, \sum_{O \in C} O, \sum_{O \in C} O^t O)$.

To store C leaders requires $O(Cd)$ space; In the case of BIRCH, the entire CF-tree needs to be stored. The CF-tree, a B^+ tree, is an abstract representation of the data; it occupies less space than the pattern matrix when the data set is large and dense clusters are present in the data. In the CF-tree, each leaf node stores a fixed number of clusters; each cluster is represented by its CF entry. Further, each cluster is made up of points that are in a sphere of pre-specified size. Each non-leaf node in the CF-tree contains a fixed number of entries, one entry for each of its children nodes. A non-leaf node may be viewed as representing a cluster made up of all the subclusters represented by its children nodes. In BIRCH, clustering is carried out in an agglomerative manner incrementally. For each point in the data set, the closest leaf node is identified by traversing the current tree structure till the leaf. If the new point fits into the leaf node based on pre-specified parameter values, it is inserted into the closest leaf. Otherwise the closest leaf node is split into two nodes appropriately. The path from the root to the leaf/leaves is updated; the corresponding CF entries are suitably changed to reflect the addition of the new point. Note that updating of the CF entries is easy because the CF vector of a non-leaf node is the sum of the CF vectors of all its children nodes. An example CF-tree is given in Figure 7.4. This figure depicts a 3-level CF-tree. There are four leaf nodes which correspond to three non-leaf nodes. These non-leaf nodes are children of a non-leaf node, the root node. The entire tree corresponds to eight clusters that have a radius less than or equal to a pre-specified threshold.

Some of the important features of BIRCH are:

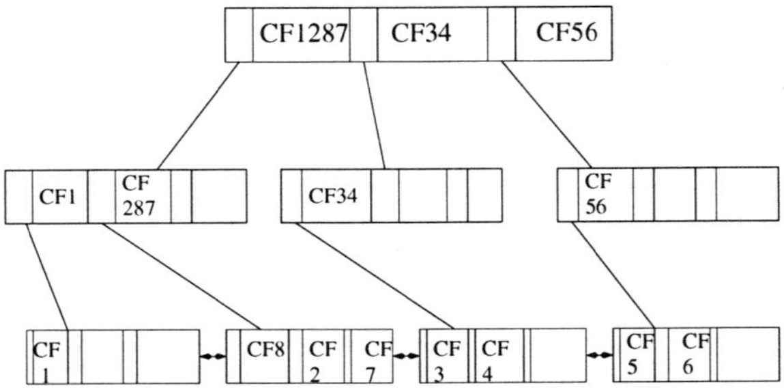

Fig. 7.4. An example CF-tree

1. It requires only one scan of the database. This is because each data item is examined only once.
2. As the resulting CF-tree structure is an abstraction of the given data, it is possible to use it for pattern synthesis or re-sampling.
3. It may be possible to use the tree structure for other data mining tasks including classification (13).
4. The tree structure can be compact and so it occupies lesser space; this may permit us to store it in the main memory.
5. The CF-tree requires $O(d)$ time and space, where d is the number of features. Note that other data structures like the KD-tree and quad-tree are exponential in the dimensionality d in both time and space.

7.2.2 Partitional Clustering

Partitional clustering algorithms generate a hard or a soft partition of the data. The most popular of this category of algorithms is the *K-Means* algorithm. A simple description of the algorithm is given below:

1. Select K out of the given n patterns as the initial cluster centers. Assign each of the remaining $n - K$ patterns to one of the K clusters; a pattern is assigned to its closest center/cluster.
2. Compute the cluster centers based on the current assignment of patterns.
3. Assign each of the n patterns to its closest center/cluster.
4. If there is no change in the assignment of patterns to clusters during two successive iterations, then stop; else, goto step 2.

There are a variety of schemes for selecting the initial cluster centers; these range from selecting the first K of the given n patterns to viewing the initial cluster seed selection as an optimization problem and using a GA to search for the globally optimal solution (23). Selecting the initial cluster centers is a very important issue. This may be illustrated with the help of the two-dimensional data set of 7 points shown in Figure 7.5 and Figure 7.6. In both the figures the same collection of points is used and it is clustered to generate a 3-partition ($K = 3$) in each case. The patterns are located at $(1,1)^t, (1,2)^t, (2,2)^t, (6,2)^t, (7,2)^t, (6,6)^t, (7,6)^t$ and are labeled A, B, C, D, E, F, G respectively. If A, D, and F are selected as the initial centers as shown in Figure 7.5, then a visually appealing partition of three clusters $\{A, B, C\}$, $\{D, E\}$, $\{F, G\}$ is generated. On the other hand, by starting with A, B, and C as the initial centers, we end up with the partition shown in Figure 7.6 which seems to have smaller variances in two clusters and a large variance in the third. Note that the variance in each cluster is acceptable in the partition shown in Figure 7.5.

An important property of the K-Means algorithm is that it implicitly minimizes the sum of squared deviations of patterns in a cluster from the center. More formally, if C_i is the ith cluster and m_i is its center, then the criterion function minimized by the algorithm is

$$\sum_{i=1}^{K} \sum_{x \in C_i} (x - m_i)^t (x - m_i).$$

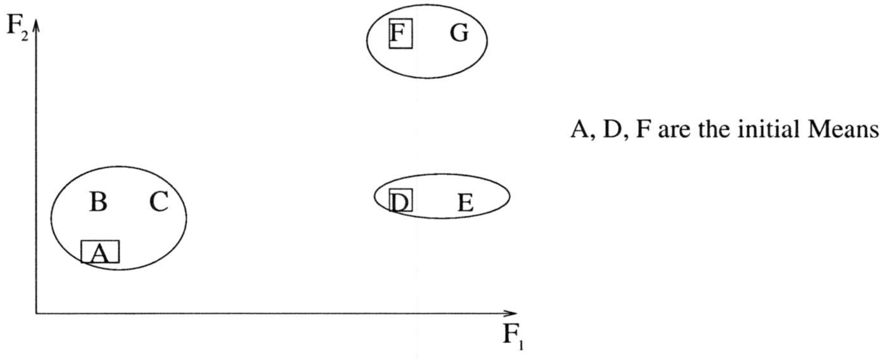

Fig. 7.5. Optimal partition

In clustering literature, it is called *sum-of-squared-error criterion,with-in-group-error sum-of-squares*, or simply *squared-error criterion*. Note that the value of the function is 2.44 for the partition in Figure 7.5, whereas its value is 17.5 for the partition in Figure 7.6. This shows that the K-Means algorithm is not guaranteed to find the globally optimal partition. A frequently used heuristic is to select the initial K centers so that they are as far away from each other as possible; this scheme is observed to work well in practice. For example, the selection made in Figure 7.5 is one such. Note that selection of A, E, and G as initial centers results in the same partition as the one shown in Figure 7.5.

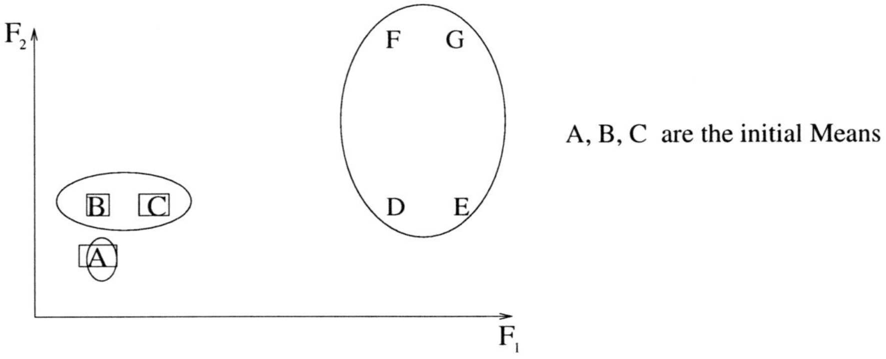

Fig. 7.6. A non-optimal partition

The time complexity of the algorithm is $O(nKdl)$, where l is the number of iterations and d is the dimensionality. The space requirement is $O(Kd)$. These features make the algorithm very attractive. It is one of the most frequently used algorithms in a variety of applications; some of these applications involve large volumes of data, for example, the satellite image data. It may not find the globally optimal partitions. Several schemes have been used to find the globally optimal partition corresponding to the minimum value of the squared-error criterion. These include (18) simulated annealing, tabu search, randomized branch and bound, and evolutionary algorithms. We focus on evolutionary algorithms in this chapter.

Evolutionary Approaches for Clustering

Evolutionary approaches, motivated by natural evolution, make use of evolutionary operators and a population of solutions to obtain the globally optimal partition of the data. The most commonly used evolutionary operators are: selection, recombination, and mutation (5; 6; 11). The most well-known evolutionary algorithms are genetic algorithms (GAs), evolution strategies (ESs), and evolutionary programming (EP). Out of these three approaches GAs have been most frequently used in clustering. Typically, solutions are binary strings in GAs. In GAs, selection operator reproduces solutions (chromosomes) to the next population based on their fitness. Selection employs a probabilistic scheme so that highly fit solutions have a higher probability of getting reproduced.

There are a variety of recombination operators in use; *crossover* is the most popular. Crossover takes as input a pair of chromosomes (called parents) and outputs a new pair of chromosomes (called children or offspring). It exchanges the segments of the parents across a crossover point. This may be illustrated using Table 7.1. Here, a single point crossover operation is depicted. The crossover point is between the fourth and the fifth bit positions. The segments in the parents after the crossover point are exchanged to produce the child chromosomes.

Table 7.1. Crossover

parent1: 1 0 1 1	0 1 0 1	child1: 1 0 1 1 1 1 1 0
parent2: 1 1 0 0	1 1 1 0	child2: 1 1 0 0 0 1 0 1

Mutation takes as input a chromosome and outputs a chromosome by complementing the bit value at a randomly selected location in the input chromosome. For example, the string '11111110' is generated by applying the mutation operator to the second bit location in the string '10111110' (starting at the left). Both crossover and mutation are applied with some pre-specified probabilities; probability of crossover (P_c) is typically large and probability of mutation (P_μ) is small. Large values of P_μ may lead to random search. Typically, evolutionary algorithms are divided into genetic algorithms (GAs), evolution strategies (ESs), and evolutionary programming (EP). GAs represent points in the search space as binary strings. GAs depend on the crossover operator to explore the search space and mutation is used in GAs for the sake of completeness, that is to make sure that no part of the search space is left unexplored. ESs and EP differ from GAs in solution representation and type of the mutation operator used. EP does not use a recombination operator. EP uses only selection and mutation operators. All these three approaches have been used to solve the clustering problem by viewing it as minimization of the squared-error criterion.

GAs search the space of possible solutions. They perform globalized search whereas others perform a localized search. Here, by localized search we mean the next solution is a neighbor, or is in the vicinity, of the current solution. Statistical algorithms like the k-means algorithm, fuzzy clustering algorithms,expectation maximization, ANNs used for clustering, various annealing schemes, and tabu search are all localized search techniques. In all these cases, there is no significant gap or difference between solutions obtained in two successive iterations. In the case of GAs the crossover and mutation operators can produce totally distinct solutions from the current ones. GAs are used to minimize the squared-error criterion. Here, each point or chromosome represents a partition of n objects into k clusters and is represented by a K-ary string of length n. The major problem with GAs is that they are sensitive to the selection of various parameters like the population size, crossover and mutation probabilities, etc. Some general guidelines for selecting these control parameters were suggested. However, these guidelines may not be adequate to get good results on specific problems like pattern clustering.

It is possible to view the clustering problem as an optimization problem that finds out the optimal centroid of the clusters directly rather than finding an optimal partition using a GA. This view permits the use of ESs and EP because centroids can be coded easily in both these approaches as they support direct-representation of a solution as a real-valued vector. It has been observed that they perform better than their classical counterparts, k-means algorithm and the fuzzy k-means algorithm. However, these approaches all suffer like GAs and ANNs from sensitivity to control parameter selection. For each specific problem, one has to tune and trim the parameter values to suit the application.

7.3 GA-Based Clustering

Genetic algorithms have been used to solve different clustering problems. Here, we focus on the application of GAs to partitional clustering. Cluster representatives, or prototypes, are used for efficient classification. These prototypes are obtained using GAs. In (6), a GA is used to get an optimal subset of prototypes. Specifically, Partitioning Around Medoids (PAM) (25) is use to get medoids of a pre-specified number of clusters present in each of the classes. The GA is used to get an optimal threshold value for each class and medoids that fall within the threshold distance from already selected medoids are eliminated. This scheme helped in reducing the number of medoids by 50% with an insignificant reduction in the classification accuracy. Similarly GAs are used in selecting a prototype set from a collection of leaders (26).

We give below a high-level description of GA-base clustering.

A Genetic Algorithm for Clustering

1. Choose a random population of solutions. Each solution here corresponds to a valid k-partition of the data. Associate a fitness value with each solution. Typically, fitness is inversely proportional to the square-error value. Larger the fitness value of a solution if its square-error value is smaller.
2. Use the evolutionary operators selection, recombination and mutation to generate the next population of solutions. Evaluate the fitness values of these solutions.
3. Repeat step 2 until some termination condition is satisfied. On termination, present the best chromosome.

The most popularly used criterion function that is used by GAs for clustering is the with-in-group-error-sum-of-squares which is associated with the k-means algorithm. In order to use a GA, the following issues have to be addressed.

1. Initialization of the population: this requires choosing the size of the population and the representation of possible solutions (strings/chromosomes) in the population. The population size is a parameter that varies form problem to problem. Typically, solutions are represented as either binary strings or as sequences of real numbers (7).
2. Compute the fitness associated with each string in the population. In the case of partitional clustering algorithms, fitness of a string is inversely proportional to the squared error value associated with the corresponding partition.
3. Genetic operators along with the associated probabilities, if any, need to be specified. These include selection, crossover, and mutation along with specification of the values for P_c and P_μ. Additional operators could be specified and used in case of need. For example, in (8), k-means operator is specified and used.
4. Termination condition: even though theoretical studies assume that GAs run over infinite iterations, in practice a threshold on the number of iterations is specified and used to terminate the GA.

We give below details on each of the issues mentioned above.

7.3.1 Initial Population

Each population is a collection of a pre-specified number (*population size*) of possible solution strings. Typically, elements of the initial population are randomly chosen. Each solution string characterizes a k-partition of the data, where k is the number of clusters of the given set of n patterns. One popular scheme for representing the solutions is to use a string of length n and allow each allele in the chromosome to take values from $\{1, 2, \cdots, K\}$. Here, each allele corresponds to a pattern and its value represents the cluster to which the pattern belongs. This kind of representation is *string-of-group-numbers* encoding (27). For example, the 2-partition $\{x_1, x_2, x_4\}$, $\{x_3, x_5\}$ of a set of 5 patterns x_1, x_2, x_3, x_4, x_5 is represented by the string 1 1 2 1 2. This representation is popular and is used in the early work (27; 28) on the topic. However, a major difficulty with this representation is that each solution string requires $O(n)$ space. So, it is not attractive for clustering a large set of patterns.

A very convenient representation is based on using a string of Kd real numbers to represent the K-partition (28). Here, each cluster is represented by its centroid as a d-dimensional vector, where d is the dimensionality. So, K clusters are represented by K centroids, one for each cluster, and Kd real numbers. For example, consider three clusters of two patterns each given by: Cluster 1- (1,1) and (2,2); Cluster 2 - (6,1) and (6,2); and Cluster 3 - (6,6) and (6,7). The centroids of the clusters respectively are (1.5, 1.5); (6.0, 1.5); and (6.0, 6.5). So, the centroid based representation of the solution string is '1.5 1.5 6.0 1.5 6.0 6.5'. This representation requires $O(Kd)$ space and is attractive to cluster large data sets when the values of K and d are small. Most of the data mining applications fall in this category; this includes clustering and classification of protein sequences.

Another popular method is to represent a chromosome as a generalized list or a tree; such a scheme is employed in genetic programming (29). A GA-based clustering scheme for dealing with large data sets is proposed and used in (4); here a hyper-quadtree is used to represent each chromosome. Each non-leaf node can have 2^d children; this property restricts it to work on low-dimensional data sets. However, it finds good solutions for large simulated low-dimensional data sets. Here, we propose a genetic algorithm based on BIRCH (22); the proposed algorithm employs CF-tree to represent each chromosome. Each node in the CF-tree requires $O(d)$ space to represent the corresponding CF vector. It has a height of $O(logn)$. These features make it attractive to deal with large high-dimensional data sets.

7.3.2 Fitness Computation

Each chromosome corresponds to a K-partition; it has an associated fitness function. The squared-error (SE) criterion corresponding to the partition is computed and the fitness is $\frac{1}{SE}$, where SE is the squared-error value. It requires $O(nd)$ time to compute the centroids of a given partition in the case of string-of-group-numbers encoding. In the case of vector of centroids, it requires $O(nKd)$ time to generate the K-partition and $O(nd)$ time to get the centroids. Using the centroids, it requires

$O(nd)$ time to get the SE value (28). The effort to compute the SE value is linear in n, K, and d. However, it requires 2 to 3 data set scans for computation of the SE value of each chromosome; it is possible to compute the fitness of all the solutions in a population in 2 to 3 data set scans. However, running the GA for l iterations means, $O(l)$ data set scans. This is the most important reason for GAs being not popular in data mining (1) where large data sets are routinely processed. In (4), a hyper-quadtree corresponding to the data set represents each chromosome; each node in the tree contains zero or more genes and initially each chromosome contains K genes and a gene is represented by its centroid. Here, an abstraction of the data set in the form of the quadtree is used for efficient processing. We propose in the next section, a scheme based on CF-tree which is a better abstraction than the quadtree for the current application.

7.3.3 Genetic Operators

Traditionally selection, crossover, and mutation are the operators used in GAs. In addition special operators for K-means clustering are used. We discuss them in detail below:

Selection

One of the prominent features of an evolutionary algorithm is the notion of *survival of the fittest*. The selection operator implements this idea by exploiting the fitness landscape; chromosomes with a higher fitness value have a larger probability of getting selected to the next population. The probability distribution is characterized by $P(s_i) = \frac{F(s_i)}{\sum_j F(s_j)}$, where $F(s_i)$ is the fitness value of string s_i. There are different schemes available for implementing the random selection scheme; of these, the roulette wheel scheme (24) is used frequently in GA-based clustering because of its simplicity. In the quadtree-based GA (4), selection is carried out independently for each subpopulation; each subpopulation is a collection of chromosomes, in the population, that have the same number of genes. So, selection from subpopulations ensures diversity in terms of number of clusters. In order to obtain fitness of a chromosome, which is inversely related to the SE value, appropriate scaling and transformation of the SE value is required. Scaled fitness value increases as the number of clusters increases, thereby favoring chromosomes with more genes. In order to control this bias, selection is carried out in subpopulations in (4).

Crossover

Crossover is a recombination operator; it produces two children from two given parents. This operator helps in exploring the search space faster. Also, it is supported by the *building block hypothesis* (24). There are a variety of crossover operators; however, the single-point crossover with a fixed value of crossover probability, P_c, is popular because of its simplicity. It is possible that crossover applied on some parent strings may result in illegal strings. For example, if the parent strings are '1 2 2 1' and '2 1 1 1', then it is possible that the children are '1 1 1 1' and '2 2 2 1'; the parents correspond to 2-partitions whereas the first child '1 1 1 1' corresponds to a single cluster. In (4), to perform crossover on two parents, a single random node in the tree is chosen and the two subtrees rooted at that node in the two parents are exchanged. This means both the parents and children have the same tree structure as all other chromosomes. The number of genes in the children could be different from that of the parents. However, crossover does not create new genes; so, the number of genes is preserved across populations.

Mutation

Mutation is a unary operator; it takes a chromosome as input and returns the chromosome with changes at different randomly selected positions. A position in the string is selected for mutation with a probability, P_μ. In (8), each allele corresponds to a data point and its value is the cluster to which the data point belongs. Here, the probability of changing the current value at a location to a new value is based on the nearness of the point to the centroid of the cluster represented by the new value; the probability is more if the point is closer to the new value. In (4), when a gene mutates, it is removed and replaced by another gene in the same chromosome. The new gene is placed in a random node and is given a value of a randomly selected point in the gene's hyperbox.

Other Operators

1. **K-means operator**: In (8) and (28), the current partitions are used to go through a K-means step which means updation of the centroids or the partition once. This operator helps in faster convergence.
2. **Replacement**: It is shown in (5) that the GA with *elitist* strategy converges to the global optimum; so, GA-based clustering algorithms use elitism, that means they copy the best strings in the current population to the next.

7.4 Clustering Large Data Sets

A major problem with the GA-based clustering algorithms presented in the previous section is that the crossover operator is not focused; for example in the centroid based representation, random subsets of centers are exchanged during crossover. This feature of the crossover operation does not help in focusing the search around promising building blocks (24) to pass on good subsets of centers from the parents to offspring. Further, these algorithms may not scale up well with the size of the data set. We describe two algorithms in this section that are suited to work on large data sets. One of them is based on hyper-quadtrees (4) and it employs the k-means clustering algorithm which requires time and space linear in the size of the data. This is described in the next subsection. Subsequently, we consider an algorithm based on BIRCH (22); it employs the CF-tree. This algorithm requires one database scan to generate the CF-tree structure from the data and is highly scalable both in terms of the number of patterns and features.

7.4.1 A Genetic Algorithm Using Hyper-Quadtrees for Low-Dimensional K-means Clustering

Hyper-Quadtrees

The quadtree is a spatial tree (for two dimensional data), where each node is associated with an axis-parallel rectangle (9). Root node is associated with an axis-parallel bounding rectangle, that tightly encloses all data points in the plane. This bounding rectangle is subdivided into four equal sized subrectangles by two axis-parallel lines passing through the rectangle's center. The root node has four children. Every child is associated with one of the four equal subrectangles. In general, every nonleaf node has four children. Each bounding rectangle associated with nonleaf node is subdivided into four equal subrectangles. This top-down construction process ends when it finds rectangles that satisfy specific termination criteria. Thus the bounding rectangles associated with the leaf nodes satisfy termination criterion. Generally, termination criterion is based on the number of data points contained in the corresponding bounding rectangle. If the number of data points in a bounding rectangle is greater than a predefined number, then further subdivide that rectangle, else make the corresponding node a leaf node.
The three-dimensional analogue of the quadtree, is called an octree. In octree, each nonleaf node has eight children and each node is associated with a axis-parallel bounding box. Each box corresponding to nonleaf node is subdivided into eight equal sized subboxes by three axis-parallel planes passing through and crossing at the box's center. The d-dimensional analogue is called multi-dimensional quadtree or hyper-quadtree. In d-dimensional hyper-quadtree, each nonleaf node has 2^d children and each node is associated with a axis-parallel bounding hyperbox. Each hyperbox corresponding to nonleaf node is subdivided by d hyperplanes passing through the hyperbox's center.
Figure 7.7(a) shows 2-dimensional data set and Figure 7.7(b) shows planar subdivision induced by quadtree.

Genetic Algorithm

Genetic algorithm for k-means clustering using hyper-quadtree is given below.

Description of each phase is given below.
Given d-dimensional data points, construct a hyper-quadtree $\mathcal{T}$. Use the following termination criterion: Do not further divide the hyperbox, if it contains less than or equal to C data points, where C is the predefined cutoff value. Use the hyper-quadtree $\mathcal{T}$ as structure of the chromosome. We have to construct the hyper-quadtree $\mathcal{T}$ only once during preprocessing. For each node in the tree $\mathcal{T}$ store the needed information

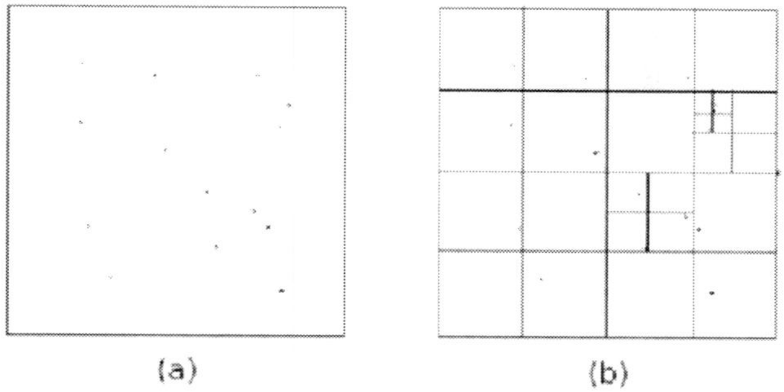

Fig. 7.7. Planar subdivision of 2-dimensional data set using quadtree

Algorithm 1 Genetic algorithm using hyper-quadtree

Require: d-dimensional data set
  Using given data set construct hyper-quadtree $\mathcal{T}$
  Initialization: Create the initial population of size μ and find fitness value for each chromosome.
  for generation 1 to N **do**
    Selection: Select μ chromosomes based on the fitness value.
    Crossover: Make pairs randomly, perform crossover to generate offspring
    Mutation: Mutate offspring
    Replacement: Consider chromosomes from the previous generation and the current generation. Out of these 2μ chromosome form a next generation of approximate size μ
  end for

like information about children to traverse the tree downward, the corresponding bounding hyperbox and center of that hyperbox.
During the running of the algorithm, several times we have to select a random node from the tree. Selection of a random node is done by using 2. To select a random node, first generate a random path starting from the root of the tree and select the node where the path terminates. Procedure for generating a random path is as follows: Start with the root of the tree and proceed towards leaf nodes. Terminate the path at current node with probability p_p, where p_p is the random path termination probability. Thus, generate a random number and if random number is less than p_p then stop the procedure and select the current node. Else randomly select one of the children nodes as current node with probability of selection proportional to the number of data points in the corresponding hyperbox. If the current node is a leaf node then stop the procedure and select that leaf node. Note that for smaller values of p_p, we get longer paths and for larger values of p_p shorter paths.
A chromosome containing k genes is called k-chromosome. A set of all k-chromosome in a given generation is called k-subpopulation. Let μ_k be the size of k-subpopulation and μ be the size of the population, then $\sum_k \mu_k = \mu$.

Algorithm 2 Select a random node

Require: Hyper-quadtree
  Initialize current node=root node
  while current node is not a leaf node **do**
    Generate a random number n
    if $n < p_p$ **then**
      Break the while loop
    else
      Randomly select one of the children
      current node = selected child
    end if
  end while
  Select the current node

Initialization

Make μ copies of the hyper-quadtree $\mathcal{T}$. All the μ trees are similar and they are chromosomes for the initial population. For each chromosome select k nodes randomly from the respective tree using algorithm 2 and assign them a gene. Gene value for the selected node is the center of the respective hyperbox. Thus, initial population has μ chromosomes, each chromosome has k genes and the population has totally $k\mu$ genes. Each node in the chromosome tree has zero or more genes.
For finding the fitness value of a chromosome, take out all (here k) gene values (center) from the chromosome tree. Run k-means algorithm using those centers for getting the partition of the data set. Take the raw fitness value of a chromosome to be SSE (square sum error) of the partition. Repeat the procedure for each chromosome.

Selection

All the chromosome trees may not contain the same number of genes, as crossover may generate offspring with number of genes different from that in the parents. Raw fitness value, calculated in initialization step and replacement step depends on number of genes. As the number of genes increases, number of clusters in k-means algorithm increases and SSE of the partition decreases. Thus it is not fair to compare two chromosome having different number of genes based on raw fitness. Thus, do selection separately for each k-subpopulation using the following method.
Calculate of the SSE of the whole data set, that is $SSE = \sum_{i=1}^{n}(x_i - x)'(x_i - x)$, where x is the centroid of the data set. For each chromosome, subtract raw fitness value from SSE of the whole data set, call it SF. Calculate average of all such fitness values, call it ASF. Then linearly scale all scaled fitness value, such that for all k-subpopulation, the maximum scaled fitness value is f times the average fitness value ASF, where f is a predefined value. Suggested value for f is 1.8 (see (24)). For each k-subpopulation, use roulette-wheel sampling proportional to the scaled fitness value, with repetition to select μ_k k-chromosomes. Thus, select a total of μ chromosomes.

Crossover

Randomly generate $\frac{\mu_k}{2}$ pairs from μ chromosomes, without repetition. For each pair generate a random number n and if n is less than crossover probability p_c, then apply crossover using the following method. Select a random node using algorithm 2. Swap the two subtrees rooted at the selected node in the two chromosome trees and make two offspring. Here the structure of the both subtree is same, so the structure of the both offspring will be same as the hyper-quadtree $\mathcal{T}$. Thus crossover does not change the structure of the tree, but it changes distribution of genes in the participating chromosome trees. Thus, it is allowed to do crossover between two chromosome trees containing different number of genes. If the number of genes in the two subtrees rooted at the selected node are different then swapping will alter the number of genes.

Mutation

Let p_m be the per chromosome mutation probability, that is the probability that a chromosome mutates in a given generation. Then for a k-chromosome, per-gene mutation rate will be:

$$p_g(k) = 1 - e^{\frac{ln(1-p_m)}{k}}.$$

Generate a random number n for each gene in each chromosome. If n is less than $p_g(k)$ then mutate the gene by following method: Remove the gene and select a random node from the chromosome tree using algorithm 2. Assign a new gene to the randomly selected node.

Replacement

Merge the parent population and offspring population, which we got after mutation. Group all chromosomes into k-subpopulations. For each k-subpopulation, order the chromosomes according to their fitness value. Top 50 percent chromosomes will go to next generation. If μ_k is odd then the chromosome, which has median fitness value, will go to next generation with 50 percent probability.

7.4.2 A Genetic Algorithm Using CF-tree for High-dimensional K-means Clustering

Section 4.1 describes use of hyper-quadtree as a structure for representing chromosomes in the genetic algorithm. For high-dimensional data such an approach is inefficient, because as the dimensionality increases, the number of children of each non-leaf node and hence the size of the tree increase exponentially. Instead we can use the CF-tree to represent chromosomes. The number of children of a node in the CF-tree is not directly dependent on the dimensionality of the data. Thus, for large dimensional data sets, one can use the algorithm described in Section 4.1 with CF-tree instead of the hyper-quadtree. But for large data sets, space required for storing

μ copies of a CF-tree or hyper-quadtree is considerably large. Thus for large data sets or a given memory size, we can use following genetic algorithm, which employs only one CF-tree.

Initialization

Given a large-dimensional large data set, construct a CF-tree $\mathcal{T}$. Use only this CF-tree for all the operations. Generate μ chromosomes, each containing k genes using the following method: For each gene in each chromosome, select a random node from CF-tree $\mathcal{T}$ and take its center value as gene. Thus, we have to store only k center values as genes for each chromosome instead of the tree itself.
For finding the raw fitness value use following method: Take out CF-vectors from the leaf nodes of the CF-tree $\mathcal{T}$. Calculate centers of all CF-vectors and make a new data set containing centers as data points. For each chromosome, run the k-means algorithm using genes as centers to make a partition of the new data set and calculate ASSE of the partition using the following equation.

$$ASSE = \sum_{j=1}^{k} \sum_{i=1}^{m_j} w_i (x_i - x_j)'(x_i - x_j),$$

where m_j is the number of data points in the jth block, w_i is the weight equal to the number of data points in the corresponding CF-vector for the data point x_i. and x_j is j^{th} center in chromosome.
One can get the partition of the original data set instead of new data set. But for large data sets time taken by proposed method is much lesser and ASSE approximates SSE. Number of data points in the new data set is much smaller compared to original data set while clustering large data sets.

Selection

Selection procedure is the same as described in Section 7.4.1. Calculate scaled fitness value for each chromosome as described in Section 7.4.1. For each k-subpopulation, use roulette-wheel sampling, proportional to scaled fitness, with repetition to select μ_k chromosomes.

Crossover

Given μ chromosomes, generate $\frac{\mu}{2}$ pairs randomly, without repetition. For each pair generate a random number n and if n is less than crossover probability p_c then apply crossover using following method: Select a random node from a CF-tree $\mathcal{T}$. Calculate center and radius of the selected node. For each gene in each chromosome, calculate the distance of the gene from the center of the selected node and mark the gene if distance is less than radius of the selected node. Swap marked genes between both chromosomes. If the number of marked genes is different in the two chromosomes, then swapping will alter the number of genes.

Mutation

Given a per-chromosome mutation probability p_m, find the per-gene mutation probability using following equation:

$$p_g(k) = 1 - e^{\frac{ln(1-p_m)}{k}}.$$

for each chromosome.
For each gene in each chromosome generate a random number n and if n is less than mutation probability $p_g(k)$ then apply mutation using following method: Select a random node from the CFtree $\mathcal{T}$ and calculate the center of the selected node. Exchange current gene center by center of the selected node.

Replacement

Replacement procedure is same as described in Section 7.4.1. Merge the parent population and offspring population. For each k-subpopulation select the best 50% chromosomes into the next population.

7.4.3 Comparison of time and space complexity

Table 7.2. Comparison of time complexity

	HQTBA	OCFTBA
Initialization	$O(\mu(S+kP))$	$O(\mu(kP))$
Selection	$O(\mu(S+k+\mu))$	$O(\mu(k+\mu))$
Crossover	$O(\mu P)$	$O(\mu(P+k))$
One Mutation	$O(P)$	$O(P)$
Replacement	$O(S+\mu nkd)$	$O(\mu n_c kd)$

Let size of the hyper-quadtree be $O(s)$ and size of the CFtree be $O(T)$ and selecting a random node requires $O(P)$ time. Creating μ copies of a hyper-quadtree requires $O(\mu S)$ time. For each chromosome, selecting k genes will require $O(kP)$ time. Hence initialization for the HQTBA (hyper-quadtree based approach) requires $O(\mu(S + kP))$ time, whereas that for the OCFTBA (one-CFtree based approach) requires $O(\mu kP)$ time because it is not needed to create μ copies.
In selection, generating all k-subpopulations requires $O(\mu)$ time in both the cases. For each k-subpopulation selecting μ_k chromosomes require $O(\mu^2)$ time. Copying each chromosome to new population requires $O(S+k)$ time in the HQTBA, whereas it requires $O(k)$ time in the OCFTBA, because we have to copy only k centers and not the whole tree. Hence selection for HQTBA requires $O(\mu(S + k + \mu))$ time, whereas that for OCFTBA requires $O(\mu(k + \mu))$ time.

In crossover, for each pair, selecting a random node requires $O(P)$ time. Swapping for HQTBA requires $O(1)$ time, whereas that for the OCFTBA requires $O(k)$ time. Because in HQTBA we just have to swap the pointers to the subtree, whereas in OCFTBA, for each gene we have to check whether the distance to the center is less than radius or not and then we have to swap marked genes. To mutate a single gene we have to select a random node, which will require $O(P)$ time for both the approaches.

In replacement, for HQTBA first we have to collect genes from the chromosome tree, which will require $O(S)$ time. Collection of genes is not required in OCFTBA. Let n be the number of total data points and n_c is the number of CF-vectors in the leaf nodes of CFtree. Each k-means iteration for the HQTBA requires $O(nkd)$ time, whereas that for OCFTBA requires $O(n_ckd)$ time. Hence replacement for HQTBA requires $O(S + \mu nkd)$ time for a population, whereas that for OCFTBA requires $O(\mu n_ckd)$ time. Clearly the time required for k-means operation dominates the running time of algorithm in both the cases and time required for k-means in OCFTBA is considerably lesser than that required for HQTBA, when $n_c << n$. Table 7.2 summarizes the time complexity of all GA operations.

Space required by the HQTBA is $O(\mu S)$, whereas that required for OCFTBA is $O(T + \mu k)$.

7.5 Multi-objective Clustering

Clustering algorithms optimize a criterion function either implicitly or explicitly. Conventionally, partitional algorithms are used to optimize a single measure of cluster quality. For example, the K-means algorithm optimizes the squared error criterion. A major problem with such an approach is that it looks for a specific structure in the data. For example, K-means algorithm is suited for extracting spherical clusters from the data. However, it can fail on data sets where the clusters are not spherical. So, the conventional clustering algorithms are not robust to variations in size and shape of clusters and variations in dimensionality and other characteristics of data. This is because they look at just one aspect of cluster quality. Two different solutions are offered in the literature to address this problem. Both are based on using **multiple objective criterion functions**. These two directions are:

1. Use different algorithms on the data set to generate a collection of partitions and obtain a consensus clustering from these (30). A related approach is to use a hybrid clustering algorithm (12) on the data set; here, different clustering algorithms are applied sequentially on the data set. Hybrid clustering helps in realizing a scalable framework for clustering.
2. Optimize different objectives explicitly in one clustering algorithm; this approach helps in exploiting an appropriate combination of the criteria explicitly. VIENNA (10) is a good example of multi-objective evolutionary algorithm for clustering. It employs specialized initialization and mutation operators. A good survey on GA-based multi-objective optimization technique is presented in (32).

An algorithm suitable for parallel processing to exploit multi-objective optimization is given in (31).

Out of these two approaches evolutionary algorithms can be judiciously combined in the second approach. Clustering is used in (32) for multi-objective optimization. It is termed clustering Pareto evolutionary algorithm (CPEA) and is useful in retaining many local Pareto-optimal frontiers thus providing an option for the designer to consider multiple potential solutions. VIENNA employs an elitist multi-objective evolutionary algorithm and is designed to provide equal opportunities for reproduction to all regions of elite nature. As a consequence, in the clustering applications, it can provide a diverse set of solutions trading off different clustering measures. The important steps in VIENNA may be summarized as follows:

1. The initial population is chosen based on Voronoi cell genotype coding. This scheme helps in selecting and employing diverse and high quality clustering solutions. Initially each element of the population is obtained by using $2k$ cluster centers which are separated from each other as far as possible.
2. It employs a directed mutation operator (and no crossover) in exploring the search space. It exploits the information of the nearest neighbors computed beforehand. mutation affects a gene and a fixed number of its neighboring data items. As a consequence, this directed mutation operator enables large changes to result from a single mutation, yet constrains them to respect local neighborhood property.
3. Given a candidate partitioning of the data, numerous measures for estimating its quality exist; some of them are variance of the resulting clusters and their separation. The two measures employed are **overall deviation** which is the sum of deviations (square of deviation is variance) and **connectivity** which evaluates the degree to which neighboring data points are placed in the same cluster.

It is shown to outperform both the K-means and the average-link agglomerative clustering algorithms on a variety of data sets of size up to 3498 patterns and 34 dimensions.

7.6 Conclusions

The role of evolutionary algorithms (EAs) in clustering is discussed in this chapter. The most popular clustering algorithm that is exploited by the EA community has been the K-means algorithm (KMA). KMA is used because of its linear time and space complexities. Most popularly used EA has been the genetic algorithms (GAs) in this context. However, most of the GA-based clustering algorithms have been used to cluster data sets with small number of patterns and/or features. The hyper-quadtree based algorithm (4) offers a faster algorithm and is used on data sets of size upto 10,000; however, because of the limitations of the hyper-quadtrees, it is not suited for high-dimensional problems. We propose a GA-based algorithm, OCFTBA, employing the CF-tree data structure. It scales up well with both the number of patterns and features. One of the comprehensive approaches to clustering based on GAs is in formulating the clustering problem as a multi-objective optimization problem and using the GA to solve it. This approach has been used on solving small or medium size clustering problems. It will be interesting to see it being applied on large-scale clustering problems.

References

[1] Mitra S, Acharya T (2003) Data mining: Multimedia, soft computing, and bioinformatics, New York: John Wiley
[2] Watanabe S (1969) Knowing and guessing, John Wiley & Sons, Inc, New York
[3] Pal S K (1996) Genetic algorithms for pattern recognition, CRC Press
[4] Freitas A A (2002) Data mining and knowledge discovery with evolutionary algorithms, Springer-Verlag, Berlin
[5] Mitchell M (1998) An introduction to genetic algorithms, Prentice-Hall of India
[6] Vose M D (2004) The simple genetic algorithm, Prentice-Hall of India
[7] Spath H (1980) Cluster analysis - algorithms for data reduction and classification of objects, Ellis Horwood, West Sussex, UK
[8] Michalewicz Z (1992) Genetic algorithms + data structures = evolution programs, Springer, NY
[9] Koza J R (1992) Genetic programming: on the programming of computers by means of natural selection. MIT Press, Cambridge, MA, USA
[10] Samet H (1990) The design and analysis of spatial data structures. Addison-Wesley Longman Publishing Co., Inc., Boston, MA, USA
[11] Goldberg D E (1989) Genetic algorithms in search, optimization, and machine learning, Addison-Wesley, Reading
[12] Jain A K, Murty M N, Flynn P J (1999) Data clustering: A review. ACM Computing Surveys 31:264-323
[13] Yu H, Yang J, Han J, Li X (2005) Making SVMs scalable to large data sets using hierarchical cluster indexing. Data Mining and Knowledge Discovery 11(3):295–321
[14] Fogel D B (1994) An introduction to simulated evolutionary optimization. IEEE Trans. Neural Networks 5(1):3-14
[15] Rudolph G (1994) Convergence analysis of canonical genetic algorithms. IEEE Trans. Neural Networks 5(1):96-101
[16] Raghavan V V, Birchard K (1979) A clustering strategy based on a formalism of the reproductive process in natural systems. SIGIR Forum, 14:10-22
[17] Jones D R, Beltramo M A (1990) Clustering with genetic algorithms. GMR-7156, General Motors Research Report
[18] Bhuyan J N, Raghavan V V, Venkatesh K E (1991) Genetic algorithm for clustering with an ordered representation. In proceedings of the Fourth ICGA, 408-415

[19] Mishra S K, Raghavan V V (1994) An empirical study of the performance of heuristic methods for clustering, in *Pattern Recognition in Practice IV*, (eds) E. S. Gelsema and L. N. Kanal, Elsevier Science, 425-436
[20] Fogel D B, Simpson P K (1993) Experiments with evolving fuzzy clusters. Proceedings of the Second Annual Conference on Evolutionary Programming
[21] Babu G P, Murty M N (1994) Clustering with evolutionary strategies. Pattern Recognition 27 (2):321-329
[22] Laszlo M, Mukherjee S (2006) A genetic algorithm using hyper-quadtrees for low-dimensional k-means clustering. IEEE Trans. Pattern Analysis Machine Intelligence 28(4):533-543
[23] Zhang T, Ramakrishnan R, Livny M (1996) BIRCH: an efficient data clustering method for very large databases. In SIGMOD-96: Proceedings of the 1996 ACM SIGMOD international conference on Management of data, 103-114, New York, NY, USA, ACM Press
[24] Babu G P, Murty M N (1993) A near-optimal initial seed selection in k-means algorithm using a genetic algorithm. Pattern Recognition Letters 14:763-769
[25] Babu T R, Murty M N (2001) Comparison of genetic algorithm based prototype selection schemes. Pattern Recognition 34:523-525
[26] Kaufman L, Rousseeuw P J (1989) Finding groups in data - An introuction to cluster analysis, Wiley, NY
[27] Krishna K, Murty M N (1999) Genetic k-means algorithm. IEEE Trans. SMC-Part B 29:433-439
[28] Murthy C A, Chowdhury N (1996) In search of optimal clusters using genetic algorithms. Pattern Recognition Letters 17:825-832
[29] Maulik U, Bandyopadhyay S (2000) Genetic algorithm-based clustering technique. Pattern Recognition 33:1455-1465
[30] Handl J, Knowles J (2004) Evolutionary multiobjective clustering. In Proceedings of PPSN VIII, LNCS 3242:1081-1091
[31] Coello Coello A C (2000) An updated survey of GA-based multiobjective optimization techniques. ACM Computing Surveys 32:109-143
[32] Hiroyasu T, Miki M, Watanabe S (2000) The new model of parallel genetic algorithm in multiobjective optimization problems- divide range multiobjective genetic algorithm. In Proceedings of the 2000 Congress on Evolutionary Computation, 333-340
[33] Molyneaux A K, Leyland G B, Favrat D (2000) A new clustering evolutionary multiobjective optimization technique, citeseer.ist.psu.edu, 446943.html

Printed in the United States
116960LV00003B/340-393/P

9 783540 774662